T0210729

A Practical Guide on Behaviour Change Support
for Self-Managing Chronic Disease

Mara Pereira Guerreiro • Isa Brito Félix
Marta Moreira Marques

Editors

A Practical Guide on Behaviour Change Support for Self-Managing Chronic Disease

 Springer

Editors

Mara Pereira Guerreiro
Nursing Research, Innovation and
Development Centre of Lisbon (CIDNUR)
Nursing School of Lisbon
Lisbon, Portugal

Isa Brito Félix
Nursing Research, Innovation and
Development Centre of Lisbon (CIDNUR)
Nursing School of Lisbon
Lisbon, Portugal

Egas Moniz Interdisciplinary Research
Center (CiiEM)
Egas Moniz School of Health & Science
Monte de Caparica, Portugal

Marta Moreira Marques
Comprehensive Health Research Centre
(CHRC) & NMS Research
NOVA Medical School, NOVA University
Lisbon, Portugal

This book is an open access publication.

ISBN 978-3-031-20009-0   ISBN 978-3-031-20010-6   (eBook)
https://doi.org/10.1007/978-3-031-20010-6

© The Editor(s) (if applicable) and The Author(s) 2023
Open Access This book is licensed under the terms of the Creative Commons Attribution 4.0
International License (http://creativecommons.org/licenses/by/4.0/), which permits use, sharing,
adaptation, distribution and reproduction in any medium or format, as long as you give appropriate credit
to the original author(s) and the source, provide a link to the Creative Commons license and indicate if
changes were made.
The images or other third party material in this book are included in the book's Creative Commons
license, unless indicated otherwise in a credit line to the material. If material is not included in the book's
Creative Commons license and your intended use is not permitted by statutory regulation or exceeds the
permitted use, you will need to obtain permission directly from the copyright holder.
The use of general descriptive names, registered names, trademarks, service marks, etc. in this publication
does not imply, even in the absence of a specific statement, that such names are exempt from the relevant
protective laws and regulations and therefore free for general use.
The publisher, the authors, and the editors are safe to assume that the advice and information in this book
are believed to be true and accurate at the date of publication. Neither the publisher nor the authors or the
editors give a warranty, expressed or implied, with respect to the material contained herein or for any
errors or omissions that may have been made. The publisher remains neutral with regard to jurisdictional
claims in published maps and institutional affiliations.

This Springer imprint is published by the registered company Springer Nature Switzerland AG
The registered company address is: Gewerbestrasse 11, 6330 Cham, Switzerland

Foreword

Due to changing patterns of disease, chronic conditions are increasingly common and present a distinct set of challenges for those offering support. Healthcare professionals have considerable education and training in providing therapies and cures, but to date, they have received less guidance on how to assist an individual in dealing with an ongoing condition. The competences required to provide professional support for individuals managing a chronic disease have been defined in the curriculum Train4Health https://www.train4health.eu/, based on a rigorous programme of work conducted by experts from across Europe. This book builds on that curriculum to offer guidance on acquiring these competences.

The authors guide the reader through background theory before introducing methods of assessing the needs of individuals with different conditions and the behaviours that need to change for optimal quality of life. Behaviour change may be necessary to manage symptoms, to adhere to medication or exercise regimens, to maintain health, and to avoid exacerbating the condition. Different behaviours may be of prime importance for different conditions, for example, changes in nutritional behaviours may be most relevant for diabetes, while smoking cessation may be most important in respiratory conditions. By following the assessment plan for each condition, the professional can identify the most important behavioural targets and then continue to subsequent chapters to learn how an intervention might be developed for individuals with each condition. The importance of understanding the barriers and facilitators is emphasised as a basis for selecting the behaviour change methods to be used. Different methods are needed if the person lacks the ability to perform the necessary behaviours than if they are not motivated to change or if they are motivated but do not succeed in making the changes. Clear guidance is offered on how to develop an intervention plan that takes account of the clinical condition, the target behaviour, barriers and facilitators, the time available, and the personal choices and resources of the individual. The step-by-step outline introduces the techniques that might be used to change behaviour and how they might be delivered to the individual. Another chapter addresses the issue of communication style, a key topic as the manner of communication about self-management may be different from the style appropriate to other facets of disease management. The guidance emphasises the importance of enabling the person to manage the changes rather than instructing or directing them and of working with them to choose the behavioural targets and the behaviour change methods.

Throughout the book, best evidence is used for each stage and references provided for the reader who wishes to read further. The editors offer a selection of current best methods of assessing behaviours, choosing behaviour change techniques, etc., allowing the reader to explore, select and adapt the approaches most suited to a specific condition and situation. Practical guidance is offered while allowing the user flexibility in how they follow the steps proposed, how they use these steps for particular applications, and the extent to which they read further about theory, assessments, and additional evidence.

This book represents a valuable step forward in ensuring that people with chronic conditions receive the best professional support in engaging in the activities needed to manage their condition successfully. Importantly, it will enable healthcare professionals to extend their competence in dealing with chronic conditions, by building on their existing therapeutic competences to gain competences in working with individuals to optimise self-management.

Professor Emeritus of Health Psychology   Marie Johnston
at the University of Aberdeen
Aberdeen, UK
June 2022

Preface

There is a story behind every book, seldom known to those who read it or use it for learning. The very beginning of the story behind this book dates to 2014, when I collaborated in an educational interdisciplinary project on virtual humans for simulation in pharmacy. The Virtual Pharmacy project opened exciting avenues of exploration in digital education.

In 2018, Isa Félix and I drew on behavioural science to develop a prototype of a mobile application with a virtual human coach to support self-management of diabetes type 2 in older people, as part of another interdisciplinary project (VASelfCare).

From this experience, we realised there was substantial room for optimising self-management support provided by health and other professionals. Brainstorming, needs assessments, pivoting and countless hours of work led us to a successful grant from the Erasmus+ programme of the European Union in 2019. The idea underpinning the Train4Health project is simple yet seemingly unexplored: combining digital education, behavioural science and co-creation with stakeholders to improve students' competencies in supporting self-management behaviours in persons with chronic disease.

Marta Marques has lent her expertise in behavioural science and health psychology to the Train4Health project since its inception. Isa, Marta, and I come from different scientific backgrounds and disciplines; we look at problems from different angles and approach work processes in different ways. We are, however, united by common values and a desire to learn from and with each other. This combination turned hard work into a rich, enjoyable and very productive experience, as it should be.

As explained in more detail in the Foreword, this book is structured in a way intending to logically lead the reader through the Train4Health innovation journey and facilitate learning. Chapter 1 introduces the competencies necessary for delivering effective behaviour change support and the corresponding learning outcomes. Chapter 2 summarises concepts and theories for understanding human behaviour. Chapter 3 moves to self-management behaviours in chronic disease and their assessment. From here, Chap. 4 discusses the assessment of behaviours determinants in practice and the implementation of behaviour change strategies. Chapter 5 recaps person-centred communication to support behaviour change. Finally, Chap. 6 points out additional learning resources developed as part of the project, including case study toolkits, and the open access simulation software.

We believe that the uniqueness of this book lies in its contribution to the development of behaviour change support competencies, through the interweave between the book's practical approach and supplementary online resources.

Ultimately, the book's contribution to the Train4Health project motto "Pushing the boundaries of behaviour change support education" depends on its permeation among academic educators and students. The strong interdisciplinary genesis of this book is expected to foster a common ground across professions and to bridge the gap between behavioural science and the education of those embracing the noble mission of helping persons with chronic disease to live healthier and happier lives.

We wish you a transformational learning experience!

Lisbon, Portugal   Mara Pereira Guerreiro
                                                                           Isa Brito Félix
                                                                     Marta Moreira Marques

Introduction

The Nursing School of Lisbon (ESEL), one of the main public higher education institutions in nursing in Portugal, has a national and international reputation. ESEL's mission is to provide undergraduate and graduate education in nursing, following the highest standards of quality grounded on the best scientific evidence. ESEL keeps up with the ever-changing nature of the world by using knowledge gained through experience and by looking beyond, to be part of and to co-create the present and future. Faithful to the values of innovation and excellence, ESEL supports research and projects that develop knowledge in nursing and contribute to the best clinical and educational practices in the health area. One example is the Train4Health project, funded by the Erasmus+ programme of the European Union, in which ESEL is the leading consortium partner.

A key challenge in the education of students is preparing them for *collaborative practice* in the health system. The seminal World Health Organization (WHO) document *"Framework for Action on Interprofessional Education & Collaborative Practice"* (WHO, 2010) established important guidelines on this matter. According to WHO (2010, p. 7), *interprofessional education occurs when students from two or more professions learn about, from and with each other to enable effective collaboration and improve health outcomes.* Effective interprofessional education enables an effective *collaborative practice,* which *happens when multiple health workers from different professional backgrounds work together with patients, families, carers and communities to deliver the highest quality of care.*

Interprofessional knowledge, practice and recognition should begin in the undergraduate education of students, which presents unique opportunities to shape future healthcare teams and to create bridges in practice. Shaping effective interdisciplinary teams takes time, and we are missing opportunities offered by undergraduate education. A team is not the sum of the expertise of different professionals; instead, it is a body with different limbs, functions and tasks to coordinately reach a common goal: better health for persons. These persons rightly expect more than fragmented care; professionals need to consider the whole person and the role of other team members to improve the quality of care.

The Train4Health project purports to push the boundaries of behaviour change support education in chronic disease through an interdisciplinary approach, contributing to a shared vision among different disciplines and setting the stage to future interprofessional education. This is a commendable endeavour in an area pivotal for patient empowerment and the sustainability of health systems.

May the motto *learning together to work together for better health* be a source of inspiration to work collaboratively and achieve excellence in healthcare.

Vice President, Nursing School of Lisbon   Patrícia Pereira
(ESEL), Train4Health Team Member
Lisbon, Portugal

The Train4Health project has received funding from the Erasmus+ Programme of the European Union under the grant agreement no. 2019-1-PT01-KA203-061389.
The European Commission's support for the production of this publication does not constitute an endorsement of the contents, which reflect the views only of the authors, and the Commission cannot be held responsible for any use which may be made of the information contained therein.

Acknowledgements

We are grateful to the Erasmus+ programme of the European Union for funding the advancement of behaviour change support education and bringing together European countries for joint work.

We would like to thank the contributors of this book for their commitment and for pursuing a common and practical vision on behaviour change support for self-managing chronic disease.

We would also like to thank the larger Train4Health team for their contribution to the project work programme.

Thank you to the consortium partners who allowed the book to be available open access: Nursing School of Lisbon, also through the Nursing Research, Innovation and Development Centre of Lisbon (CIDNUR), Inholland University of Applied Sciences, Royal College of Surgeons in Ireland, Sports Sciences School of Rio Maior – Polytechnic Institute of Santarém, the University of Lisbon, and the University of Maribor.

Finally, we greatly appreciated Springer's collaboration throughout the journey that made this book possible.

Contents

Contributors

Cristina Lavareda Baixinho Nursing Research, Innovation and Development Centre of Lisbon (CIDNUR), Nursing School of Lisbon, Lisbon, Portugal

Katja Braam Faculty of Health, Sports and Social Work, Inholland University of Applied Sciences, Haarlem, Netherlands

José Camolas Serviço de Endocrinologia, Centro Hospitalar Universitário Lisboa Norte, Lisbon, Portugal

Egas Moniz Interdisciplinary Research Center (CiiEM), Egas Moniz School of Health & Science, Egas Moniz University, Almada, Portugal

Laboratório de Nutrição, Faculdade de Medicina Universidade de Lisboa, Lisbon, Portugal

Maria Beatriz Carmo LASIGE, Faculdade de Ciências, Universidade de Lisboa, Lisbon, Portugal

Afonso Miguel Cavaco Faculty of Pharmacy University of Lisbon, Lisbon, Portugal

Ana Paula Cláudio LASIGE, Faculdade de Ciências, Universidade de Lisboa, Lisbon, Portugal

Luís Correia LASIGE, Faculdade de Ciências, Universidade de Lisboa, Lisbon, Portugal

David de Sousa Loura Centro Hospitalar Universitário de Lisboa Central, Hospital Dona Estefânia, Lisbon, Portugal

Nursing Research, Innovation and Development Centre of Lisbon (CIDNUR), Nursing School of Lisbon, Lisbon, Portugal

Isa Brito Félix Nursing Research, Innovation and Development Centre of Lisbon (CIDNUR), Nursing School of Lisbon, Lisbon, Portugal

Lucija Gosak Faculty of Health Sciences, University of Maribor, Maribor, Slovenia

Mara Pereira Guerreiro Nursing Research, Innovation and Development Centre of Lisbon (CIDNUR), Nursing School of Lisbon, Lisbon, Portugal

Egas Moniz Interdisciplinary Research Center (CiiEM), Egas Moniz School of Health & Science, Egas Moniz University, Monte de Caparica, Portugal

Helga Rafael Henriques Nursing Research, Innovation and Development Centre of Lisbon (CIDNUR), Nursing School of Lisbon, Lisbon, Portugal

Maria Adriana Henriques Nursing Research, Innovation and Development Centre of Lisbon (CIDNUR), Nursing School of Lisbon, Lisbon, Portugal

Afke Kerkstra Inholland University of Applied Sciences, Amsterdam, The Netherlands

Mateja Lorber Faculty of Health Sciences, University of Maribor, Maribor, Slovenia

Anabela Mendes Department of Medical-Surgical Nursing Adult/Elderly, Nursing School of Lisbon, Lisbon, Portugal

Nursing Research, Innovation and Development Centre of Lisbon (CIDNUR), Nursing School of Lisbon, Lisbon, Portugal

Marta Moreira Marques Comprehensive Health Research Centre (CHRC) & NMS Research, NOVA Medical School, NOVA University, Lisbon, Portugal

Nuno Pimenta Sport Sciences School of Rio Maior – Polytechnic Institute of Santarém, Rio Maior, Portugal

Interdisciplinary Centre for the Study of Human Performance, Faculty of Human Kinetics, University of Lisbon, Lisbon, Portugal

Centro de Investigação Interdisciplinar em Saúde, Instituto de Ciências da Saúde, Universidade Católica Portuguesa, Lisbon, Portugal

Carlos Filipe Quitério Centro Hospitalar de Setúbal, Setúbal, Portugal

Gregor Štiglic Faculty of Health Sciences, University of Maribor, Maribor, Slovenia

Faculty of Electrical Engineering and Computer Science, University of Maribor, Maribor, Slovenia

Usher Institute, University of Edinburgh, Edinburgh, UK

Judith Strawbridge School of Pharmacy and Biomolecular Sciences, Royal College of Surgeons in Ireland, Dublin, Ireland

Abbreviations

A1C	Glycated Haemoglobin Levels
ADA	American Diabetes Association
AFHC	Adolescents' Food Habit Checklist
AUDIT	Alcohol Use Disorders Identification Test
BMI	Body Mass Index
BCTTv1	BCT Taxonomy version 1
BCTs	Behaviour Change Techniques
COPD	Chronic Obstructive Pulmonary Disease
DASH	Dietary Approaches to Stop Hypertension
FCQ	Food Choice Questionnaire
GINA	Global Initiative for Asthma
GNKQ	General Nutrition Knowledge Questionnaire
GOLD	Global Initiative for Chronic Obstructive Lung Disease
HAPA	Health Action Process Approach
HBM	Health Belief Model
HF	Heart Failure
IDF	International Diabetes Federation
MoD	Modes of Delivery
MOOC	Massive Open Online Course
SDM	Shared Decision-Making
SDT	Self-Determination Theory
SCT	Social Cognitive Theory
SUS	System Usability Scale
TPB	Theory of Planned Behaviour
T2D	Type 2 Diabetes
WHO	World Health Organization

List of Figures

List of Tables

List of Boxes

Chapter 1
Interprofessional Learning Outcomes-Based Curriculum to Support Behaviour Change in Persons Self-Managing Chronic Disease

Afke Kerkstra, Cristina Lavareda Baixinho, Isa Brito Félix, Judith Strawbridge, and Mara Pereira Guerreiro

Chronic diseases, also known as non-communicable diseases, are a global epidemic, responsible for the majority of deaths worldwide. In the European region, chronic diseases account for 89.6% of deaths (World Health Organization, 2017). Cardiovascular diseases, cancers, diabetes and chronic lung diseases present the highest prevalence, which is projected to increase in the coming years (World Health Organization, 2017).

In addition to morbidity and mortality, chronic disease leads to marked costs for governments and individuals, as well as losses for economies. The economic burden of chronic disease includes direct costs, both medical (e.g. medication, hospital stays, ambulatory consultations) and non-medical (e.g. transportation for healthcare services), plus indirect costs (e.g. productivity loss of the person and/or caregiver,

Supplementary Information: The online version contains supplementary material available at https://doi.org/10.1007/978-3-031-20010-6_1.

A. Kerkstra (✉)
InHolland University of Applied Sciences, Amsterdam, The Netherlands
e-mail: afke.kerkstra@inholland.nl

C. L. Baixinho · I. B. Félix
Nursing Research, Innovation and Development Centre of Lisbon (CIDNUR),
Nursing School of Lisbon, Lisbon, Portugal

J. Strawbridge
School of Pharmacy and Biomolecular Sciences, Royal College of Surgeons in Ireland,
Dublin, Ireland

M. P. Guerreiro
Nursing Research, Innovation and Development Centre of Lisbon (CIDNUR),
Nursing School of Lisbon, Lisbon, Portugal

Egas Moniz Interdisciplinary Research Center (CiiEM), Egas Moniz School
of Health & Science, Monte de Caparica, Portugal

© The Author(s) 2023
M. P. Guerreiro et al. (eds.), *A Practical Guide on Behaviour Change Support for Self-Managing Chronic Disease*, https://doi.org/10.1007/978-3-031-20010-6_1

workforce attrition, early retirement). For example, health expenditure for adults with diabetes aged 20–79 years old, irrespective of being borne by individuals, their families or the health system, grew globally from USD 232 billion in 2007 to USD 966 billion in 2021. This represents a 316% increase over 15 years (International Diabetes Federation, 2021).

Addressing the burden of chronic diseases has been a priority for the European Commission. The most recent example is the Healthier Together – European Union Non-Communicable Diseases initiative (2022), which reinforces the support for action of the Member States and relevant stakeholders in this area in five strands: cardiovascular diseases, diabetes, chronic respiratory diseases, mental health, neurological disorders and health determinants.

Self-management is defined as tasks performed by an individual to minimise the impact of one's disease, with or without the support of health professionals (Lorig & Holman, 2003). Tasks can holistically be categorised under medical management (e.g. taking medication, adhering to a diet, engaging in physical activity), role management (e.g. redefining life roles in light of chronic disease) and emotional management (e.g. dealing with anger and frustration) and are related to a set of skills (Lorig & Holman, 2003). Self-management is a key component of managing chronic diseases. Taking diabetes as an example, it has been estimated that persons with this condition spend fewer than 6 hours per year consulting with healthcare professionals (Holt & Speight, 2017). It is therefore critical to empower persons to deal with chronic diseases on a daily basis.

Health and other professionals are expected to support self-management and, in particular, behaviour change. Evidence highlights that competent behaviour change counselling is still regarded as uncommon in clinical practice, suggesting a global healthcare problem. Meta-research, including 36 systematic reviews, showed that healthcare professionals frequently miss opportunities to provide brief behaviour change advice, even when it is perceived as needed (Keyworth et al., 2020). Training on behaviour change interventions to support self-management is not always part of professional training (Keyworth et al., 2019).

A scoping review found that behaviour change techniques (BCTs) remain underused in self-management interventions (Riegel et al., 2021). One reason might be the poor permeation of behavioural science, and BCTs in particular, into the education and training of health and other professionals. Difficulties in providing effective support of self-management behaviours in chronic disease are also compounded by the lack of a common set of knowledge and skills in health and other professionals.

The Train4Health project (https://www.train4health.eu) is a strategic partnership involving seven European institutions across five countries, which seeks to improve behaviour change support competencies to support self-management in chronic disease. The project target groups are nursing, pharmacy and sport sciences students, due to their pivotal role in supporting self-management of persons living with chronic diseases. Nonetheless, behaviour change support education is of interest to a wider range of professions, including dentists, nutritionists, psychologists, physicians and physiotherapists.

Train4Health envisages a continuum in behaviour change support education, in which an interprofessional competency framework, relevant for those currently practising, guided the development of a learning outcomes-based curriculum and an educational package for future professionals (today's undergraduate students). The next section addresses the development of the competency framework.

1.1 The Interprofessional Train4Health Competency Framework

The development of a European competency framework for health and other professionals to support behaviour change in persons self-managing chronic disease has been detailed elsewhere (Guerreiro et al., 2021).

In essence, the framework comprises 26 competency statements, classified into 2 categories, depicted in Fig. 1.1. Competencies that directly support behaviour change in the self-management of chronic disease (BC1–BC14) are presented in Fig. 1.2; foundational competencies required for effective delivery of behaviour change support (F1–F12) are described in the competency framework paper (Guerreiro et al., 2021).

Competency statements were consensualised through an e-Delphi panel, composed of 48 participants from 12 European countries and a variety of disciplines: pharmacy, nursing, sport sciences, physiotherapy, general practice, nutrition, psychology and public health (Guerreiro et al., 2021).

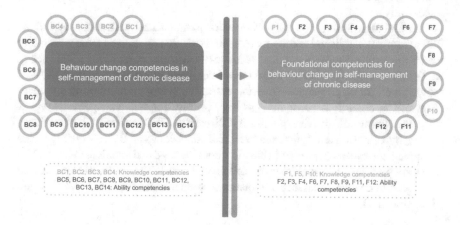

Fig. 1.1 European competency framework for health and other professionals to support behaviour change in persons self-managing chronic disease

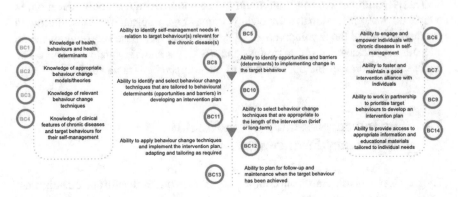

Fig. 1.2 Behaviour change competencies

The central part of Fig. 1.2 depicts the traditional assessment–planning–intervention–monitoring cycle, familiar to health and other professionals, from a behaviour change support perspective. BC1, BC2, BC3 and BC4, depicted on the left side of Fig. 1.2, indicate knowledge required to deliver behaviour change support in chronic disease. On the right side, BC6, BC7, BC9 and BC14 are communication and relationship abilities, essential across the assessment–planning–intervention–monitoring cycle.

General communication competencies and professionalism are encompassed in the foundational category (e.g. F6, ability to communicate effectively in partnership with people and families, and F11, ability to demonstrate professional behaviour, respectively).

The Train4Health competency framework is associated with a core set of 21 BCTs from an established taxonomy (Michie et al., 2013), derived from a literature search in conjunction with experts' feedback (Guerreiro et al., 2021). This set of BCTs can be employed in five behaviours (diet, including alcohol intake, physical activity, medication adherence, smoking cessation, symptom monitoring and management), in conditions recognised as high priority for self-management: type 2 diabetes, chronic obstructive pulmonary disease (COPD), obesity, heart failure, asthma, hypertension and ischaemic heart disease. Additional BCTs were organised in supplementary sets per target behaviour; both the core and supplementary lists of BCTs are presented as supplementary online material 1. These lists of standardised techniques to change behaviour are linked to competencies BC3, BC10, BC11 and BC12 (Fig. 1.2).

1.2 From Behaviour Change Competencies to a Learning Outcomes-Based Curriculum

The Train4Health European competency framework was the starting point for developing the learning outcomes-based curriculum, which has been detailed elsewhere (Cadogan et al., 2021). Two principles guided curriculum development, the interprofessional nature of behaviour change support and the importance of inter-professional education, in which different professionals are brought together to learn with, from and about one another (Reeves et al., 2016).

Typically, in education, the term "curriculum" refers to a set of components in a course: learning outcomes, teaching strategies, student activities and assessments, which should be aligned (Kennedy, 2006; Cedefop, 2017). The designation "learning outcomes-based curriculum" stresses the pivotal role of learning outcomes in determining content to be included in the Train4Health educational products.

Firstly, information on behaviour change curricula was collated at the partner institutions as part of needs assessment. This was analysed and supplemented with a literature search to develop a master list of learning outcomes (Loura et al., 2021). This exercise served to further highlight the learning outcomes that were generally covered in curricula and gaps according to the competency framework.

A draft list of learning outcomes was then compiled for each competency statement. The learning outcomes were written in accordance with best practice (Kennedy, 2006; Cedefop, 2017) using the style "The learner is (or will be) able to", followed by an action verb, so that students are able to demonstrate what they have learned. Different verbs were used to demonstrate different levels of learning in accordance with Bloom's taxonomy of learning (Bloom et al., 1964; Kennedy, 2006).

Content was determined by working backwards from the learning outcomes associated with behaviour change competencies. Those related to foundational competencies were deemed as more comprehensively addressed by existing curricula and therefore not developed as content. Each learning outcome was numbered in relation to a competency statement. Learning outcomes and associated content were iteratively refined, as depicted in Fig. 1.3.

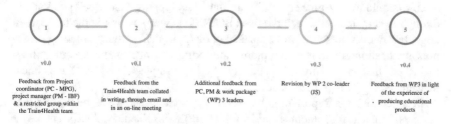

Fig. 1.3 Iterative improvement of the Train4Health learning outcomes and curriculum content

A total of 23 pre-essential learning outcomes were developed, associated with foundational competencies, and 34 learning outcomes associated with behaviour change competencies: 12 related to knowledge and 22 to ability. Within the latter, selected learning outcomes are presented in Chaps. 2, 3, 4 and 5, alongside with their respective content. Supplementary online material 2 presents the current version of the learning outcomes associated with behaviour change competencies, including Bloom's taxonomy level, illustrating different levels of complexity (cognitive and affective), and the proposed curriculum content.

The domains of competency and the learning outcomes developed provide an overview of the knowledge, skills and attitudes needed by healthcare graduates (Pontefract & Wilson, 2019) to effectively support behaviour change for the self-management of chronic disease.

1.3 Practical Considerations About the Learning Outcomes-Based Curriculum

We believe that the work presented here is the first attempt to develop a transnational competency framework and related curriculum on behaviour change support in chronic disease for health and other professionals. This work is expected to promote consistency of required competencies and learning outcomes across countries and higher education institutions throughout Europe. Furthermore, it may raise awareness about the potential need for curricular transformation and pave the way for benchmarking and the identification of best practices.

Using standardised BCTs in behaviour change support education is expected to facilitate tailored interventions while fostering comprehensiveness and consistency. Another envisaged benefit is enabling a clearer description of behaviour change support in practice. Specifically, training future professionals in using BCTs responds to the quest of facilitating design, reporting and comparability of self-management interventions, put forward by researchers that developed and validated a taxonomy of self-management interventions for chronic conditions (Orrego et al., 2021). In fact, the approach presented in this book extends the scope of Orrego et al.'s taxonomy, by further specifying the "1.1 Support technique" subdomain, through the explicit use of standardised BCTs. Also worth noting is the fit between target behaviours presented in Chap. 3 (e.g. diet, physical activity, smoking cessation, medication adherence, symptom monitoring and management) with subdomains of this taxonomy (respectively, "2.1 Lifestyle-related"; "2.2 Clinical management") (Orrego et al., 2021).

In education, it is important to be aware of the relation between the level of taxonomy of the learning outcomes and teaching methods (Kennedy, 2006). The lower level of cognitive learning outcomes, for example, is suitable for the Massive Open Online Course (MOOC), which is one of the Train4Health educational products. For higher-level cognitive learning outcomes, more interactive teaching methods are needed to enhance analysis, synthesis and evaluation. The Train4Health case studies, detailed in Chap. 6, are suitable for this purpose.

In a curriculum, there should be clear alignment between learning outcomes, teaching strategies, student activities and assessment tasks (Kennedy, 2006; Cedefop, 2017; Alfauzan & Tarchouna, 2017). Therefore, a critical step is the definition of links between learning outcomes, teaching strategies, student activities and assessment tasks (Kennedy, 2006). Case studies toolkits and the simulation software, presented in Chap. 6, contain assessment tasks linked with learning outcomes.

Another challenge is understanding which learning outcomes are best achieved by interprofessional learning. It is valuable to understand the nature and extent of learning outcomes that are common to all the professions (Steven et al., 2017) and where there are differences between the professions.

Behaviour change support education lends itself to interprofessional education, as competencies needed to support behaviour change in persons with chronic diseases are common across health and other professions. Interprofessional education is, however, still in a nascent stage, despite its advantages (Dyess et al., 2019; Reeves et al., 2016; Wang et al., 2019). These include improving learners' knowledge, skills and understanding of interprofessional practice (Cox et al., 2016), improving attitudes towards teamwork and collaborative practises and developing skills for interprofessional communication (Dyess et al., 2019; Wang et al., 2019).

Accomplishing interprofessional education is more challenging than using different professionals to develop resources, having different students from different professions using the same resource or showcasing the roles of the different professionals. The design of this learning outcomes-based curriculum encourages educators and students to develop teaching and learning flexibly, which is an important step towards an interprofessional approach.

The Train4Health competency framework and the corresponding curriculum may be adapted for brief, less complex interventions than behaviour change in chronic disease. These include encouraging a healthy lifestyle in general (increasing physical activity, adopting a healthy diet), promoting an active ageing and supporting medication adherence. Another transferability potential is stimulating the alignment with communication skills required to effectively apply behaviour change techniques. The novelty of the Train4Health project is supporting the training of undergraduate students, facilitating future performance and reducing workforce challenges (Vallis et al., 2017).

Key Points
- Health and other professionals are expected to support self-management of chronic disease, but education on behaviour change support is often suboptimal.
- The Train4Health framework comprises 26 competency statements, classified into 2 categories: competencies that directly support behaviour change in the self-management of chronic disease (BC1–BC14) and foundational competencies required for effective delivery behaviour change (F1–F12).
- The Train4Health competency framework is associated with a core set of 21 BCTs from an established taxonomy.
- A total of 23 pre-essential learning outcomes were developed, associated with foundational competencies, and 34 learning outcomes associated with behaviour change competencies: 12 related to knowledge and 22 to ability.

- In a curriculum, there should be clear alignment between learning outcomes, teaching strategies, student activities and assessment tasks.
- Behaviour change support education lends itself to interprofessional education, as competencies needed to support behaviour change in persons with chronic diseases are common across health and other professions.

References

Alfauzan, A. H., & Tarchouna, N. (2017). The role of an aligned curriculum design in the achievement of learning outcomes. *Journal of Education and e-Learning Research, 4*(3), 81–91. https://doi.org/10.20448/journal.509.2017.43.81.91

Bloom, B. S., Masia, B. B., & Krathwohl, D. R. (1964). *Taxonomy of educational objectives volume II: The affective domain.* McKay.

Cadogan, C., Strawbridge, J., Cavaco, A., Kerkstra, A., Baixinho, C., Félix, I., Marques, M.M., & Guerreiro, M. P. (2021). *Report on the development of a European competency framework for health and other professionals to support behaviour change in persons self-managing chronic disease and the associated common learning outcomes-based curriculum.* ISBN 978-989-53445-0-5.

Cedefop. (2017). *Defining, writing and applying learning outcomes: A European handbook.* Publications Office. https://doi.org/10.2801/566770

Cox, M., Cuff, P., Brandt, B., Reeves, S., & Zierler, B. (2016). Measuring the impact of interprofessional education on collaborative practice and patient outcomes. *Journal of Interprofessional Care, 30*(1), 1–3. https://doi.org/10.3109/13561820.2015.1111052

Dyess, A. L., Brown, J. S., Brown, N. D., Flautt, K. M., & Barnes, L. J. (2019). Impact of interprofessional education on students of the health professions: A systematic review. *Journal of Educational Evaluation for Health Professions, 16*, 33. https://doi.org/10.3352/jeehp.2019.16.33

Guerreiro, M. P., Strawbridge, J., Cavaco, A. M., Félix, I. B., Marques, M. M., & Cadogan, C. (2021). Development of a European competency framework for health and other professionals to support behaviour change in persons self-managing chronic disease. *BMC Medical Education, 21*(1), 287. https://doi.org/10.1186/s12909-021-02720-w

Healthier Together – European Union Non-Communicable Diseases initiative. (2022). https://ec.europa.eu/health/non-communicable-diseases/overview_en

Holt, R. I. G., & Speight, J. (2017). The language of diabetes: The good, the bad and the ugly. *Diabetic Medicine, 34*(11), 1495–1497. https://doi.org/10.1111/dme.13520

International Diabetes Federation. (2021). *IDF Diabetes Atlas 10th.* https://diabetesatlas.org

Kennedy, D. (2006). Writing and using learning outcomes: A practical guide. Cork: University College Cork. https://cora.ucc.ie/bitstream/handle/10468/1613/A%20Learning%20Outcomes%20Book%20D%20Kennedy.pdf?sequence=1&isAllowed=y

Keyworth, C., Epton, T., Goldthorpe, J., Calam, R., & Armitage, C. J. (2019). 'It's difficult, I think it's complicated': Health care professionals' barriers and enablers to providing opportunistic behaviour change interventions during routine medical consultations. *British Journal of Health Psychology, 24*(3), 571–592. https://doi.org/10.1111/bjhp.12368

Keyworth, C., Epton, T., Goldthorpe, J., Calam, R., & Armitage, C. J. (2020). Delivering opportunistic behavior change interventions: A systematic review of systematic reviews. *Prevention Science, 21*(3), 319–331. https://doi.org/10.1007/s11121-020-01087-6

Lorig, K., & Holman, H. (2003). Self-management education: History, definition, outcomes and mechanisms. *Annals of Behavioral Medicine, 26*(1), 1–7. https://doi.org/10.1207/S15324796ABM2601_01

Loura, D., Arriscado, A. E., Kerkstra, A., Nascimento, C., Félix, I., Guerreiro, M., & Baixinho, C. (2021). Interprofessional competency frameworks in health to inform curricula development: Integrative review. *New Trends in Qualitative Research, 6*, 63–71. https://doi.org/10.36367/ntqr.6.2021.63-71

Michie, S., Richardson, M., Johnston, M., Abraham, C., Francis, J., Hardeman, W., Eccles, M. P., Cane, J., & Wood, C. E. (2013). The behavior change technique taxonomy (v1) of 93 hierarchically clustered techniques: Building an international consensus for the reporting of behavior change interventions. *Annals of Behavioral Medicine, 46*(1), 81–95. https://doi.org/10.1007/s12160-013-9486-6

Orrego, C., Ballester, M., Heymans, M., Camus, E., Groene, O., Niño de Guzman, E., Pardo-Hernandez, H., & Sunol, R. (2021). Talking the same language on patient empowerment: Development and content validation of a taxonomy of self-management interventions for chronic conditions. *Health Expectations, 24*, 1626–1638. https://doi.org/10.1111/hex.13303

Pontefract, S. K., & Wilson, K. (2019). Using electronic patient records: Defining learning outcomes for undergraduate education. *BMC Medical Education., 19*, 30. https://doi.org/10.1186/s12909-019-1466-5

Riegel, B., Westland, H., Iovino, P., Barelds, I., Bruins Slot, J., Stawnychy, M. A., Osokpo, O., Tarbi, E., Trappenburg, J. C. A., Vellone, E., Strömberg, A., & Jaarsma, T. (2021). Characteristics of self-care interventions for patients with a chronic condition: A scoping review. *International Journal of Nursing Studies, 116*, 103713. https://doi.org/10.1016/j.ijnurstu.2020.103713

Reeves, S., Fletcher, S., Barr, H., Birch, I., Boet, S., Davies, N., McFadyen, A., Rivera, J., & Kitto, S. (2016). A BEME systematic review of the effects of interprofessional education: BEME Guide No. 39. *Medical Teacher, 38*(7), 656–668. https://doi.org/10.3109/0142159X.2016.1173663

Steven, K., Howden, S., Mires, G., Rowe, I., Lafferty, N., Arnold, A., & Strath, A. (2017). Toward interprofessional learning and education: Mapping common outcomes for prequalifying healthcare professional programs in the United Kingdom. *Medical Teacher, 39*(7), 720–744. https://doi.org/10.1080/0142159X.2017.1309372

Vallis, M., Lee-Baggley, D., Sampalli, T., Ryer, A., Ryan-Carson, S., Kumanan, K., & Edwards, L. (2017). Equipping providers with principles, knowledge and skills to successfully integrate behaviour change counselling into practice: A primary healthcare framework. *Public Health, 154*, 70–78. https://doi.org/10.1016/j.puhe.2017.10.022

Wang, Z., Feng, F., Gao, S., & Yang, J. (2019). A systematic meta-analysis of the effect of Interprofessional education on health professions students' attitudes. *Journal of Dental Education, 83*(12), 1361–1369. https://doi.org/10.21815/JDE.019.147

World Health Organization. (2017). *Non-communicable diseases*. World Health Organization. https://doi.org/10.5005/jp/books/11.

Open Access   This chapter is licensed under the terms of the Creative Commons Attribution 4.0 International License (http://creativecommons.org/licenses/by/4.0/), which permits use, sharing, adaptation, distribution and reproduction in any medium or format, as long as you give appropriate credit to the original author(s) and the source, provide a link to the Creative Commons license and indicate if changes were made.

The images or other third party material in this chapter are included in the chapter's Creative Commons license, unless indicated otherwise in a credit line to the material. If material is not included in the chapter's Creative Commons license and your intended use is not permitted by statutory regulation or exceeds the permitted use, you will need to obtain permission directly from the copyright holder.

Chapter 2
Concepts and Theories in Behaviour Change to Support Chronic Disease Self-Management

Maria Adriana Henriques and David de Sousa Loura

Learning Outcomes

This chapter contributes to achieving the following learning outcomes:

BC1.1 Differentiate between health behaviour and behaviour determinants.

BC2.1 Describe the approach of different models and theories to behaviour change in health.

BC2.2 Provide a rationale for using behaviour change models/theories.

BC2.3 Explain how different models and theories predict self-management behaviours in chronic disease and allow an understanding of interventions that can change these behaviours.

Population ageing and the increasing burden of chronic diseases require the health system to support self-management (Araújo-Soares et al., 2019). Many have argued that behaviour change support is central to the needs of persons with chronic diseases. Although modifiable health behaviours are recognised as a major influence in the management of chronic conditions, there are several barriers to effective behaviour change support by health and other professionals, such as compartmentalisation in silos of healthcare provision, lapses in communication between different teams and lack of training (Vallis et al., 2019).

Ultimately, to meet the real needs of persons living with chronic diseases, professionals need education fostering the development of behaviour change competencies,

M. A. Henriques (✉)
Nursing Research, Innovation and Development Centre of Lisbon (CIDNUR),
Nursing School of Lisbon, Lisbon, Portugal
e-mail: ahenriques@esel.pt

D. de Sousa Loura
Nursing Research, Innovation and Development Centre of Lisbon (CIDNUR),
Nursing School of Lisbon, Lisbon, Portugal

Centro Hospitalar Universitário de Lisboa Central, Hospital Dona Estefânia, Lisbon, Portugal

© The Author(s) 2023
M. P. Guerreiro et al. (eds.), *A Practical Guide on Behaviour Change Support for Self-Managing Chronic Disease*, https://doi.org/10.1007/978-3-031-20010-6_2

as described by the interprofessional competency framework presented in Chap. 1. In particular, health and other professionals need to acquire concepts and definitions from behavioural science, to better understand how to support people living with chronic diseases.

In this chapter, we address the concepts of health behaviours, the difference in relation to behaviour determinants and, briefly, key behaviour change models and theories that can inform self-management interventions in chronic disease.

2.1 Health Behaviours and Behaviour Determinants

Chronic diseases are the result of the interplay between genetic, physiological, environmental and behaviour determinants. This means that health-related decisions that people make will impact on their health outcomes. Some have even argued that, in the case of chronic conditions, health outcomes are more influenced by the choices the individual makes than by the care directly provided by health professionals (Vallis et al., 2019).

The interest in behaviours that have an impact on health and well-being is shaped, particularly, by the assumption that such behaviours are modifiable (Conner, 2002). Health behaviour has been defined in various ways, as illustrated in Table 2.1.

As opposed to acute conditions, in chronic diseases the role of health and other professionals is more focused on supporting and empowering persons to adopt and maintain behaviours, such as medication adherence, not smoking, keeping a healthy weight, being physically active, consuming substances in moderation, eating healthily and getting adequate sleep. It has long been recognised that these health-protective behaviours have a significant impact in preventing both morbidity and mortality, also impacting "(...) upon individuals' quality of life, by delaying the onset of chronic disease and extending active lifespan" (Conner, 2002). More recently, a European multicohort study in 116,043 people free of major chronic diseases at baseline, from 1991 to 2006, suggests that various healthy lifestyle profiles, such as regular physical activity, BMI less than 25, not smoking and moderate alcohol consumption, appear to yield gains in life-years without major chronic diseases, including type 2 diabetes, coronary heart disease, stroke, cancer, asthma and chronic obstructive pulmonary disease (Nyberg et al., 2020).

Table 2.1 Definitions of "health behaviour"

Authors	Health behaviour definition
Conner & Norman (1996)	"(...) any activity undertaken for the purpose of preventing or detecting disease or for improving health and well-being"
Gochman (1997)	"(...) behaviour patterns, actions and habits that relate to health maintenance, to health restoration and to health improvement"
Parkerson et al. (1993)	"(...) actions of individuals, groups, and organisations, as well as their determinants, correlates, and consequences, including social change, policy development and implementation, improved coping skills, and enhanced quality of life"

Based on Conner (2002) and Glanz et al. (2008)

It is therefore important to distinguish between health-impairing (or health-risk) and health-enhancing (or health-protective) behaviours:

"Health impairing behaviours have harmful effects on health or otherwise predispose individuals to disease. Such behaviours include smoking, excessive alcohol consumption, and high dietary fat consumption. In contrast, engagement in health enhancing behaviours convey health benefits or otherwise protect individuals from disease. Such behaviours include exercise, fruit and vegetable consumption, and condom use in response to the threat of sexually transmitted diseases." (Conner, 2002)

Table 2.2 depicts examples of health outcomes associated with the core health-risk behaviours; as already pointed out, these outcomes may be influenced by other factors than individual behaviours *per se*, such as genetics or environmental issues.

Behaviour determinants are factors that influence the behaviour, either in a positive or in a negative way, i.e. facilitators of and barriers to the performance of the behaviour, respectively. These determinants can be individual, such as biological/demographic (i.e. age, health history) and psychological factors (i.e. motivation, humour, previous experiences), or non-individual, such as aspects related to the behaviour itself (i.e. time needed to see changes), environmental/social issues (i.e. interpersonal relationships, such as social support) and health policies (i.e. policies targeting the health problem and making it easier/harder to change).

In fact, behaviour determinants can be categorised in a multitude of ways, one of which is a widely used model of behaviour change – the COM-B model (Michie et al., 2011, 2014a) – which will be further described in this chapter. This model groups determinants into three major categories: capability (physical and psychological), motivation (reflexive and automatic) and opportunity (social and physical) (Michie et al., 2011, 2014a).

Table 2.2 Relationship between health behaviours and health outcomes

Health behaviour	Health outcome
Smoking	Coronary heart disease Cancer (lung, throat, stomach and bowel) Reduced lung capacity and bronchitis
Excessive fat consumption and insufficient fibre, fruit and vegetable consumption	Cardiovascular diseases, stroke, high blood pressure, cancer, diabetes, obesity, osteoporosis, dental disease
Sedentarism	Increased cardiovascular morbidity and mortality High blood pressure Decreased metabolism of carbohydrates and fats
Non-adherence to health screenings	Undetectability of severe health problems, such as anaemia, diabetes, high blood pressure, bronchitis and cervical and breast cancer
Impaired sexual behaviours	Sexually transmitted diseases (STDs) (such as gonorrhoea, syphilis and human immunodeficiency virus)
Alcohol overconsumption	Health problems (high blood pressure, heart disease, cirrhosis of the liver) Social problems (accidents, injuries, suicides, crime, domestic violence, murder and unsafe sex)

Based on Conner (2002)

Due to the complexity of health-related human behaviour, for which usually various determinants are responsible for the adoption and sustainability of the behaviours, behavioural science has dedicated much of its research to the development, testing and refinement of models and theories that aim to explain, predict, model and change behaviour. These are described in the next section.

2.2 Behaviour Change Theories and Models

There are various definitions of theories. In behavioural science, a theory has been defined as "a set of concepts and/or statements which specify how phenomena relate to each other, providing an organising description of a system that accounts for what is known, and explains and predicts phenomena" (Davis et al., 2015, p. 327). In brief, theories are a systematic way of understanding phenomena (e.g. health behaviour(s)), in which a set of concepts and variables (e.g. determinants of behaviours such as motivation, self-efficacy and social influences) and prepositions about their relationships are established (Glanz et al., 2002).

In the context of health behaviour change, theories "(...) seek to explain why, when, and how a behaviour does or does not occur, and to identify sources of influence to be targeted in order to alter behaviour" (Michie et al., 2018, p. 70). They are important to understand disease patterns and to explain why individuals and communities have certain behaviours (Simpson, 2015), as well as to support clinical practice and professionals in day-to-day interventions. Using a theory when developing and evaluating interventions allows the identification of what needs to change (i.e. the determinants of the behaviour), the barriers and facilitators to changing those influences and the identification of mechanisms of action that operate along the pathway to change. Further, many theories identify ways in which the behaviour determinants can be modified, i.e. which techniques can be implemented in interventions to, for instance, increase motivation, increase the capability to perform the behaviour and form habits.

There is therefore consensus that health behaviour change interventions should be informed by theory. For example, the UK's Medical Research Council recommends the identification of relevant theories for designing complex interventions (Skivington et al., 2021), as does the US Department of Health and Human Services and National Institutes of Health (Rimer & Glanz, 2005). Guideline PH49 from the National Institute for Health and Care Excellence (NICE) on individual-level health behaviour change interventions recommends that staff should be offered professional development on behaviour change theories, methods and skills (NICE, 2014).

There are many theories of behaviour and behaviour change; a multidisciplinary literature review has identified 83 theories containing a total of 1725 concepts (Davis et al., 2015). A summary of these theories, their concepts and relationships are available in a theory database (Hale et al., 2020).

We present an overview of the main theories that have been applied to self-management of chronic diseases. Table 2.3 presents a brief description and key references for key social-cognitive, motivation and integrative theories (see Michie et al., 2014b, for more details about these theories). Examples of applications are presented in accompanying papers.

Table 2.3 Overview of core behaviour change models and theories in the self-management of chronic disease

Theory or model	Brief description	Key references on the theory or model	References of application of the theory or model
Health belief model	The health belief model (HBM) seeks to predict if it is likely that a person will follow the recommended action other than a health-impairing path. It conceives health behaviour as a product of two cognitions – The perception of illness threat and evaluation of behaviours to counteract this threat, through perceived susceptibility, perceived severity, perceived benefits and perceived barriers	Rosenstock (1974) Janz and Becker (1984)	Jiang et al. (2021)
Theory of planned behaviour	The theory of planned behaviour (TPB) is one of the most widely used theories in the health domain. It is an extension of an early theory, theory of reasoned action, and it describes behaviour as a product of behavioural intention, which in turn is determined by the following cognitions: Attitude (degree to which the behaviour is positively or negatively valued), subjective norm (perceived social pressure to engage or not engage in the behaviour) and perceived behavioural control (extent to which people perceive themselves as capable of performing a given behaviour). Intentions are the major precursors of the behaviour, in which attitudes are determined by behavioural beliefs, subjective norms by normative beliefs and perceived behavioural control by control beliefs. TPB is useful to predict intention to perform a behaviour but not in explaining how to move from intention to action	Ajzen (1991) Conner (2010)	Rich et al. (2015) Senkowski et al. (2019)
Social cognitive theory	Social cognitive theory (SCT) proposes that behavioural, environmental and personal factors interact to determine each other (triadic reciprocity). It describes human functioning through a number of basic capabilities: Symbolising capability, forethought capability, vicarious capability, self-regulatory capability and self-reflective capability. One of the most important constructs in SCT is self-efficacy, i.e. one's own belief or confidence in their capability to perform a given action and cope with the difficulties that may arise. SCT describes how self-efficacy is influenced, e.g. verbal persuasion and previous mastery experiences, which has guided many health behaviour change interventions	Bandura (1982) Bandura (1986)	Tougas et al. (2015) Romeo et al. (2021)

(continued)

Table 2.3 (continued)

Theory or model	Brief description	Key references on the theory or model	References of application of the theory or model
Control theory	Control theory uses engineering principles to explain how individuals select and achieve their goals through negative feedback loops (discrepancy between current state and desired goal). The theory explains how individuals exert control to deal with external or internal disturbances to goal achievement. Control theory provides information about strategies that can be used to change behaviours, e.g. self-monitoring and feedback, action control strategies and goal revision	Carver & Scheier (1982) Carver & Scheier (1998)	Prestwich et al. (2016)
Self-determination theory	Self-determination theory (SDT) is a theory of human motivation, which states that motivation varies in its quality, in a continuum from autonomous forms of motivation (such as intrinsic motivation – Enjoyment) to controlled forms of motivation (driven by external and internal pressures), to amotivation (lack of motivation). Autonomous forms of motivation are associated with sustained health behaviour changes, whereas controlled forms of motivation lead to worst health outcomes. Autonomous motivation results from the satisfaction of three basic psychological needs – Need for autonomy, relatedness and competence. To be effective, behaviour change interventions need to ensure the satisfaction of these needs, which can be achieved with the implementation of certain techniques	Deci and Ryan (1985) Ryan & Deci (2017)	Ntoumanis et al. (2021) Teixeira et al. (2020)
Health action process approach (HAPA)	HAPA model aims to explain how behaviours are developed and maintained over time, dividing behaviour change in motivational (or decision-making) and action (or maintenance) phases. In the motivational phase, people form an intention to perform a health-protective behaviour via self-efficacy, outcome expectancies and threat perception. In the action phase, operationalisation of intentions occurs, relying on self-regulatory processes to maintain the behaviour over time. Situational barriers can impair these processes and lead to behaviour maintenance failure in action performance	Schwarzer (1992)	Zhang et al. (2019) Chen et al. (2020)

Based on Michie et al. (2014a, b, 2018)

2.3 COM-B Model: A Model to Guide Interventions

Given the multiplicity of theories of behaviour that exist and the need for a simplified model to support the design of behaviour change interventions, Michie and colleagues developed the COM-B model (Michie et al., 2011), which defines behaviour as a result of the individual's capability, opportunity and motivation to follow a certain course of action. *Capability* is an attribute of a person that reflects the *psychological* and *physical* abilities to perform a behaviour, including knowledge and skills. *Motivation* is an aggregate of mental processes that energise and direct behaviour; it involves the *automatic* and *reflective* motivation, such as the individual's goals, plans and beliefs and their emotions, habits or impulses, respectively. *Opportunity* is an attribute of an environmental system, corresponding to the external factors that may facilitate or make it harder for an individual to adopt a specific behaviour, such as the *physical* and *social environment* (Michie et al., 2011; West & Michie, 2020). Table 2.4 presents definitions of the COM-B components (West & Michie, 2020).

For particular behaviours, people, context or phase of the behaviour change process, certain barriers or facilitators will have more influence than the others. The analyses of these determinants play a crucial role in selecting which strategies will work better to achieve the desired behaviour, as described in Chap. 4.

This model has been successfully used in designing behaviour change interventions (e.g. Issac et al., 2021; Wheeler et al., 2018), to support behaviour change in practice. We present an example related to behaviour change in the self-management of a chronic disease. Wheeler et al. (2018) researched the feasibility and usability of a mobile intervention for persons with hypertension, using the COM-B model as a framework for the development of a knowledge model to support the intervention. Knowing that each person has a behaviour source that relies on different factors, the intervention was tailored to the user's individual profile, based on the initial contact. Using the COM-B model, the authors identified different profiles via a

Table 2.4 Definition of COM-B model components

Capability	Motivation	Opportunity
Physical capability is capability that involves a person's physique and musculoskeletal functioning (e.g. balance and dexterity)	*Reflective motivation* is motivation that involves conscious thought processes (e.g. plans and evaluations)	*Physical opportunity* is opportunity that involves inanimate parts of the environmental system and time (e.g. financial and material resources)
Psychological capability is capability that involves a person's mental functioning (e.g. understanding and memory)	*Automatic motivation* is motivation that involves habitual, instinctive, drive-related and affective processes (e.g. desires and habits)	*Social opportunity* is opportunity that involves other people and organisations (e.g. culture and social norms)

self-evaluation questionnaire on determinants for sodium intake and physical activity. Examples of these determinants are synthesised in Table 2.5, based on Wheeler et al. (2018).

As depicted in Fig. 2.1, according to this model, all three categories of determinants influence the adoption and maintenance of health-protective behaviours, and consequently the successful self-management of chronic diseases.

Table 2.5 Examples of behaviour determinants for increasing physical activity using the COM-B model

Behaviour target: increase physical activity		
COM-B model component	Examples of behaviour determinants (expressed as facilitators of the behaviour)	Intervention
Physical capability	Physical strength and stamina	Educational and motivational messages, notifications and reminders about the behaviour
Psychological capability	Knowledge about the importance of physical activity	
Physical opportunity	Having access to suitable facilities to be physically active, e.g. safe walking path	
Social opportunity	Joining an exercise group	
Automatic motivation	Including exercise as a part of the daily routine	
Reflective motivation	Belief about the positive benefits of physical exercise to one's health	

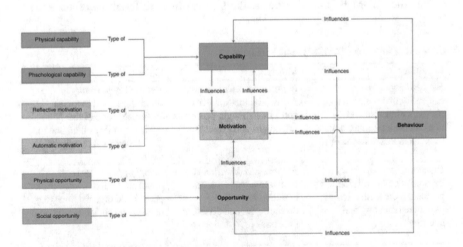

Fig. 2.1 Illustration of the COM-B model (West & Michie, 2020)

Key Points

- A health behaviour is any behaviour that can affect a person's health either by enhancing it (health-protective behaviour) or impairing it (health-damaging or health-risk behaviour).
- Behaviour determinants are individual or non-individual aspects which can influence the behaviour either in a positive or a negative way, i.e. barriers or facilitators for the adoption of the behaviour.
- Theories and models of behaviour change provide a deeper understanding about behaviour patterns, responses and their determinants; they can guide effective behaviour change support.
- The COM-B model presents an integrative approach, in which a behaviour is considered to be influenced by physical and psychological capability, reflective and automatic motivation and physical and social opportunity.

References

Ajzen, I. (1991). The theory of planned behavior. *Organizational Behavior and Human Decision Processes, 50*, 179–211. https://doi.org/10.1016/0749-5978(91)90020-T

Araújo-Soares, V., et al. (2019). Developing behavior change interventions for self-Management in Chronic Illness: An integrative overview. *European Psychologist, 24*, 7–25. https://doi.org/10.1027/1016-9040/a000330

Bandura, A. (1982). Self-efficacy mechanism in human agency. *The American Psychologist, 37*, 122–147. https://doi.org/10.1037/0003-066X.37.2.122

Bandura, A. (1986). *Social foundations of T ought and action: A social cognitive theory.* Prentice-Hall.

Carver, C. S., & Scheier, M. F. (1982). Control theory: A useful conceptual framework for personality-social, clinical and health psychology. *Psychological Bulletin, 92*(1), 111–135. https://doi.org/10.1037/0033-2909.92.1.111

Carver, C. S., & Scheier, M. F. (1998). *On the self-regulation of behaviour.* Cambridge University Press.

Chen, Y., Tan, D., Xu, Y., Wang, B., Li, X., Cai, X., Li, M., Tang, C., Wu, Y., Shu, W., Zhang, G., Huang, J., Zhang, Y., Yan, Y., Liang, X., & Yu, S. (2020). Effects of a HAPA-based multicomponent intervention to improve self-management precursors of older adults with tuberculosis: A community-based randomised controlled trial. *Patient Education and Counseling, 103*(2), 328–335. https://doi.org/10.1016/j.pec.2019.09.007

Conner, M. (2002). Health behaviors. *International Encyclopedia of the Social & Behavioral Sciences, 2015.* https://doi.org/10.1016/B978-0-08-097086-8.14154-6.

Conner, M. (2010). Cognitive determinants of health behavior. In A. Steptoe (Ed.), *Handbook of behavioral medicine* (pp. 19–30). Springer Science and Business Media. https://doi.org/10.1007/978-0-387-09488-5_2

Conner, M., & Norman, P. (Eds.). (1996). *Predicting health behaviour.* Open University Press.

Davis, R., Campbell, R., Hildon, Z., Hobbs, L., & Michie, S. (2015). Theories of behaviour and behaviour change across the social and behavioural sciences: A scoping review. *Health Psychology Review, 9*(3), 323.44. https://doi.org/10.1080/17437199.2014.941722

Deci, E. L., & Ryan, R. M. (1985). *Intrinsic motivation and self-determination in human behaviour.* Plenum Publishing.

Glanz, K., Rimer, B., & Lewis, F. (2002). *Health behavior and health education.* John Wiley & Sons, Inc.

Glanz, K., Rimer, B., & Viswanath, K. (Eds.). (2008). *Health behavior and health education: Theory, research, and practice* (4th ed.). Jossey-Bass.

Gochman, D. (1997). Health behavior research: Definitions and diversity. In D. Gochman (Ed.), *Handbook of health behavior research (Vol. I), personal and social determinants.* Plenum Press.

Hale, J., Hastings, J., West, R., Lefevre, C. E., Direito, A., Connell Bohlen, L., Godinho, C., Anderson, N., Zink, S., Goarke, H., & Michie, S. (2020). *An ontology-based modelling system (OBMS) for representing behaviour change theories applied to 76 theories [version 1; peer review: Awaiting peer review].* Wellcome Open Research.

Issac, H., Taylor, M., Moloney, C., & Lea, J. (2021). Exploring factors contributing to chronic obstructive pulmonary disease (COPD) guideline non-adherence and potential solutions in the emergency department: Interdisciplinary staff perspective. *Journal of Multidisciplinary Healthcare, 14*, 767–785. https://doi.org/10.2147/JMDH.S276702

Janz, N. K., & Becker, M. H. (1984). The health belief model: A decade later. *Health Education Quarterly, 11*(1), 1–47. https://doi.org/10.1177/109019818401100101

Jiang, L., Liu, S., Li, H., Xie, L., & Jiang, Y. (2021). The role of health beliefs in affecting patients' chronic diabetic complication screening: A path analysis based on the health belief model. *Journal of Clinical Nursing, 30*, 2948–2959. https://doi.org/10.1111/jocn.15802

Michie, S., van Stralen, M., & West, R. (2011). The behaviour change wheel: A new method for characterising and designing behaviour change interventions. *Implementation Science, 6.* https://doi.org/10.1186/1748-5908-6-42

Michie, S., Atkins, L., & West, R. (2014a). *The behaviour change wheel: A guide to designing interventions.* Silverback Publishing.

Michie, S., Campbell, R., Brown, J., West, R., & Gainforth, H. (2014b). *ABC of behaviour change theories.* Silverback Publishing.

Michie, M., Marques, M., Norris, E., & Johnston, M. (2018). Theories and interventions in health behavior change. In T. Revenson & R. Gurung (Eds.), *Handbook of health psychology.* Routledge. https://doi.org/10.4324/9781315167534

National Institute for Health and Care Excellence. (2014). NICE Guidance: Behaviour change: individual approaches. Retrieved from https://www.nice.org.uk/guidance/ph49

Ntoumanis, N., Ng, J., Prestwich, A., Quested, E., Hancox, J., Thøgersen-Ntoumani, C., Deci, E., Ryan, R., Lonsdale, C., & Williams, G. (2021). A meta-analysis of self-determination theory-informed intervention studies in the health domain: Effects on motivation, health behavior, physical, and psychological health. *Health Psychology Review, 15*(2), 214–244. https://doi.org/10.1080/17437199.2020.1718529

Nyberg, S. T., Singh-Manoux, A., Pentti, J., Madsen, I. E. H., Sabia, S., Alfredsson, L., et al. (2020). Association of healthy lifestyle with years lived without major chronic diseases. *JAMA Internal Medicine,* 1–9. https://doi.org/10.1001/jamainternmed.2020.0618

Parkerson, G., et al. (1993). Disease-specific versus generic measurement of health-related quality of life in insulin dependent diabetic patients. *Medical Care, 31*, 629–637. https://doi.org/10.1097/00005650-199307000-00005

Prestwich, A., Conner, M., Hurling, R., Ayres, K., & Morris, B. (2016). An experimental test of control theory-based interventions for physical activity. *British Journal of Health Psychology, 21*(4), 812–826. https://doi.org/10.1111/bjhp.12198

Rich, A., Brandes, K., Mullan, B., & Hagger, M. S. (2015). Theory of planned behavior and adherence in chronic illness: A meta-analysis. *Journal of Behavioral Medicine, 38*(4), 673–688. https://doi.org/10.1007/s10865-015-9644-3

Rimer, B. K., & Glanz, K. (2005). Theory at a glance: A guide for health promotion practice. Bethesda, MD: US Department of Health and Human Services, National Institutes of Health, National Cancer Institute.

Romeo, A. V., Edney, S. M., Plotnikoff, R. C., et al. (2021). Examining social-cognitive theory constructs as mediators of behaviour change in the active team smartphone physical activity program: A mediation analysis. *BMC Public Health, 21*, 88. https://doi.org/10.1186/s12889-020-10100-0

Rosenstock, I. (1974). Historical origins of the health belief model. *Health Education Monographs, 2*(4), 328–335. https://doi.org/10.1177/109019817400200403

Ryan, R. M., & Deci, E. L. (2017). *Self-determination theory: Basic psychological needs in motivation, development, and wellness*. Guilford Publishing.

Schwarzer, R. (1992). Self-efficacy in the adoption and maintenance of health behaviours: Theoretical approaches and a new model. In R. Schwarzer (Ed.), *Self-efficacy: Thought control of action* (pp. 217–243). Hemisphere.

Senkowski, V., Gannon, C., & Branscum, P. (2019). Behavior change techniques used in theory of planned behavior physical activity interventions among older adults: A systematic review. *Journal of Aging and Physical Activity, 27*(5), 746–754. https://doi.org/10.1123/japa.2018-0103

Simpson, V. (2015). *Models and theories to support health behavior intervention and program planning*. Available from: https://extension.purdue.edu/extmedia/HHS/HHS-792-W.pdf

Skivington, K., Matthews, L., Simpson, S. A., Craig, P., Baird, J., Blazeby, J. M., Boyd, K. A., Craig, N., French, D. P., McIntosh, E., Petticrew, M., Rycroft-Malone, J., White, M., & Moore, L. (2021). A new framework for developing and evaluating complex interventions: Update of Medical Research Council guidance. *BMJ* (Clinical research ed.), 374, n2061. https://doi.org/10.1136/bmj.n2061

Teixeira, P. J., Marques, M. M., Silva, M. N., Brunet, J., Duda, J. L., Haerens, L., La Guardia, J., Lindwall, M., Lonsdale, C., Markland, D., Michie, S., Moller, A. C., Ntoumanis, N., Patrick, H., Reeve, J., Ryan, R. M., Sebire, S. J., Standage, M., Vansteenkiste, M., et al. (2020). A classification of motivation and behavior change techniques used in self-determination theory-based interventions in health contexts. *Motivation Science, 6*(4), 438–455. https://doi.org/10.1037/mot0000172

Tougas, M., Hayden, J., McGrath, P., Huguet, A., & Rozario, S. (2015). A systematic review exploring the social cognitive theory of self-regulation as a framework for chronic health condition interventions. *PLoS One, 10*(8), e0134977. https://doi.org/10.1371/journal.pone.0134977

Vallis, M., et al. (2019). Integrating behaviour change counselling into chronic disease management: A square peg in a round hole? A system-level exploration in primary health care. *Public Health, 175*, 43–53. https://doi.org/10.1016/j.puhe.2019.06.009

West, R., & Michie, S. (2020). A brief introduction to the COM-B model of behaviour and the PRIME theory of motivation. *Qeios*. https://doi.org/10.32388/WW04E6

West, R., Godinho, C. A., Bohlen, L. C., Carey, R. N., Hastings, J., Lefevre, C. E., & Michie, S. (2019). Development of a formal system for representing behaviour-change theories. *Nature Human Behaviour, 3*(5), 526. https://doi.org/10.1038/s41562-019-0561-2

West, R., Michie, S., Rubin, J., & Amlôt, R. (2020). Applying principles of behaviour change to reduce SARS-CoV-2 transmission. *Nature Human Behaviour, 4*, 451–459. https://doi.org/10.1038/s41562-020-0887-9

Wheeler, T., Vallis, T., Giacomantonio, N., & Abidi, S. (2018). Feasibility and usability of an ontology-based mobile intervention for patients with hypertension. *International Journal of Medical Informatics, 119*. https://doi.org/10.1016/j.ijmedinf.2018.08.002

Zhang, C., Zhang, R., Schwarzer, R., & Hagger, M. S. (2019). A meta-analysis of the health action process approach. *Health Psychology, 38*, 623–637. https://doi.org/10.1037/hea0000728

Open Access   This chapter is licensed under the terms of the Creative Commons Attribution 4.0 International License (http://creativecommons.org/licenses/by/4.0/), which permits use, sharing, adaptation, distribution and reproduction in any medium or format, as long as you give appropriate credit to the original author(s) and the source, provide a link to the Creative Commons license and indicate if changes were made.

The images or other third party material in this chapter are included in the chapter's Creative Commons license, unless indicated otherwise in a credit line to the material. If material is not included in the chapter's Creative Commons license and your intended use is not permitted by statutory regulation or exceeds the permitted use, you will need to obtain permission directly from the copyright holder.

Chapter 3
Identifying and Assessing Self-Management Behaviours

Helga Rafael Henriques, José Camolas, Nuno Pimenta, and Anabela Mendes

Learning Outcomes

This chapter contributes to achieving the following learning outcomes:

BC1.2 Describe target behaviours in the self-management of chronic diseases.

BC4.2 Identify higher- and lower-level target behaviours for specific chronic diseases, based on appropriate guidance.

BC5.1 Assess the person's behaviour in self-management using appropriate measures.

H. R. Henriques (✉)
Nursing Research, Innovation and Development Centre of Lisbon (CIDNUR),
Nursing School of Lisbon, Lisbon, Portugal
e-mail: hrafael@esel.pt

J. Camolas
Serviço de Endocrinologia, Centro Hospitalar Universitário Lisboa Norte, Lisbon, Portugal

Centro de Investigação Interdisciplinar Egas Moniz (CiiEM), Egas Moniz University,
Almada, Portugal

Laboratório de Nutrição, Faculdade de Medicina Universidade de Lisboa, Lisbon, Portugal

N. Pimenta
Sport Sciences School of Rio Maior – Polytechnic Institute of Santarém, Rio Maior, Portugal

Interdisciplinary Centre for the Study of Human Performance, Faculty of Human Kinetics,
University of Lisbon, Lisbon, Portugal

Centro de Investigação Interdisciplinar em Saúde, Instituto de Ciências da Saúde,
Universidade Católica Portuguesa, Lisbon, Portugal

A. Mendes
Department of Medical-Surgical Nursing Adult/Elderly, Nursing School of Lisbon,
Lisbon, Portugal

Nursing Research, Innovation and Development Centre of Lisbon (CIDNUR),
Nursing School of Lisbon, Lisbon, Portugal

© The Author(s) 2023
M. P. Guerreiro et al. (eds.), *A Practical Guide on Behaviour Change Support for Self-Managing Chronic Disease*, https://doi.org/10.1007/978-3-031-20010-6_3

3.1   Target Behaviours for the Self-Management of Chronic Diseases

Seven high-priority chronic diseases were selected, based on their prevalence and potential for self-management, according to two European Union (EU)-funded projects addressing this topic (COMPAR-EU and PRO-STEP): type 2 diabetes (T2D), chronic obstructive pulmonary disease (COPD), hypertension, heart failure (HF), obesity, asthma and ischaemic heart disease. The self-management of these diseases encompasses, but is not restricted to, the change or maintenance of health-enhancing or health-protective behaviours presented in Table 3.1, called target behaviours. For listing these target behaviours, we followed the 6S hierarchy of pre-appraised evidence (DiCenso et al., 2009).[1]

Moreover, target behaviours were selected on the grounds of their overall relevance to persons living with these chronic diseases; behaviours pertinent only to specific cases (e.g. nutritional supplementation in malnourished persons with COPD) were not considered.

Symptom monitoring and management addresses physical or mental alterations associated with the chronic disease, which the person can notice. Examples include managing hypoglycaemia in diabetes or recording asthma symptoms. Other self-management components, such as self-monitoring of blood glucose when beneficial in persons with T2D, do not respect to "symptoms" and therefore are not addressed in this chapter.

The target behaviours presented above describe broad behaviours, designated as *high-level target behaviours*. These can usually be broken down into more granular behaviours, focusing on specific behavioural actions, designated as *low-level target behaviours*. For instance, physical activity, i.e. all voluntary muscle actions associated with increased energy expenditure (Caspersen et al., 1985), is a high-level target behaviour. Physical activity includes both daily living activities and exercise training and can, therefore, encompass low-level target behaviours, such as walking or swimming, as illustrated in Fig. 3.1.

A healthy diet is another high-level target behaviour. A healthy diet arises from a proper balance between its elements (foods) and their constituents (nutrients). The latter may vary in amount, depending on factors such as maturation and soil characteristics, which influence fruits' and vegetables' composition, and animal feed, which influence meats' and eggs' composition, to give just two examples. Moreover, nutrients interact with each other – e.g. dietary fibre influences cholesterol absorption and metabolism. Therefore, the same nutrient may have a different impact on

---

[1] We used the highest possible layer in the 6S model, in the case, evidence-based current clinical guidelines and other summaries of evidence (e.g. Cryer, 2019). Lower layers in this model include, in descending order, synopsis of syntheses, summarising the findings of high-quality systematic reviews, often accompanied by a commentary on the methodological quality and the clinical applicability of its findings, and syntheses (systematic reviews). The bottom layer of the 6S model are single original studies, which have not been pre-appraised.

Table 3.1 Self-management behaviours in high-priority chronic diseases

Behaviour	Type 2 diabetes (ADA, 2021)	COPD (GOLD, 2020)	Hypertension (Williams et al., 2018)	Heart failure (McDonagh et al., 2021)	Obesity (Yumuk et al., 2015)	Asthma (GINA, 2021)	Ischaemic heart disease (Knuuti et al., 2020)
Diet (including alcohol intake)	•	–	•	•	•	•	•
Physical activity	•	•	•	•	•	–	•
Medication adherence	•	•	•	•	•	•	•
Smoking cessation	•	•	•	•	–	•	•
Symptom monitoring and management	•	•	–	•	–	•	•

From Guerreiro et al. (2021)

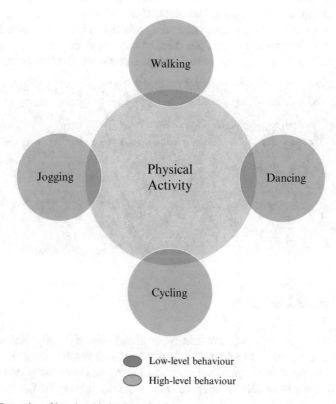

Fig. 3.1 Examples of low-level behaviours in physical activity

health, depending on the food chosen and the clinical condition of the person eating it. Supporting the adoption of a healthy diet in practice requires the definition of more granular behaviours, such as reducing the intake of free sugars or limiting salt intake to no more than 5 g/day.

The target behaviours listed in Table 3.1 are informed by a biomedical perspective. They can, however, also be defined in terms emanating from the person living with chronic disease, grounded on a set of preferences, resources and experiences, unique to that person. This is exemplified, for medication-taking, by "Being able to walk to the supermarket without breathlessness" in a person with COPD or "Adjusting insulin to preferably reduce hypoglycaemia episodes to zero" in a person with T2D managed with insulin. As professionals and persons with chronic disease work collaboratively, materialising paradigms such as shared decision-making and person-centred communication, these target behaviours, defined from the person's perspective, will predictably become more common.

In the next sections, examples of low-level target behaviours for the self-management of each high-priority chronic disease are provided. Disease management interventions relying necessarily on professionals, such as ordering laboratory tests, initiating medicines and pulmonary rehabilitation and cardiac rehabilitation programmes, are outside the remit of this book.

Evidence levels underpinning target behaviours were defined differently across sources. For details about the evidence levels, refer to the international clinical guidelines, mentioned in the footnotes across this chapter.

Box 3.1 presents websites where potential updates of cited international clinical guidelines for the management of the selected chronic disease can be checked; it may be equally relevant to resort to national guidelines, where these exist.

Box 3.1 Websites of International Clinical Guidelines: Examples

- American Diabetes Association (ADA) practice guidelines resources: https://professional.diabetes.org/content-page/practice-guidelines-resources
- European Society of Cardiology (ESC) guidelines and scientific documents: https://www.escardio.org/Guidelines
- Global initiative for chronic obstructive lung disease (GOLD): https://goldcopd.org
- Global initiative for asthma (GINA): https://ginasthma.org

3.1.1 Type 2 Diabetes

Type 2 diabetes is the most prevalent type of diabetes. Hyperglycaemia results, initially, from the inability of cells to respond fully to insulin. Insulin resistance prompts an increase in the production of this hormone; over time pancreatic beta cells are frequently unable to keep up with demand, leading to the insufficient production of insulin (IDF, 2021). The cornerstone of type 2 diabetes management is promoting a lifestyle composed of a healthy diet, regular physical activity, smoking cessation and weight loss, when needed. If lifestyle changes are insufficient to manage hyperglycaemia, pharmacologic treatment is initiated.

3.1.1.1 Diet

Diet plays a key role in the overall management of diabetes. There is no single diet pattern for persons with T2D. Research provides clarity on many food choices and eating patterns that can help these persons in achieving health goals, such as lower glycated haemoglobin levels (A1C), lower self-reported weight, lower lipid levels and quality of life (ADA, 2021). Therefore, a variety of eating behaviour patterns are acceptable for the self-management of T2D, which should be chosen based on the person's cultural and socio-economic background, preferences and needs (ADA, 2021).

Healthy eating recommendations are appropriate for persons with type 2 diabetes, when an individualised meal plan is not set. Table 3.2 presents low-level dietary target behaviours, based on the ADA guidelines (2021).

Obesity or overweight affects many persons with type 2 diabetes, reaching 90–95% in the American population (CDC, 2021). A simple manner to explain obesity physiopathology is considering that it results from a maintained positive energy balance. In other words, excessive weight arises from energy intake (food calories) continuously exceeding expenditure. For weight loss purposes, the restriction of energy intake is robustly supported by literature.

Some low-level target behaviours for self-managing type 2 diabetes are common to the general population. For example, the World Health Organization (2015) recommends a reduced intake of free sugars, for both adults and children, to less than 10% of total energy intake and suggests that it is further reduced below 5% (WHO, 2015). The term "free sugars" refers to all monosaccharides and disaccharides added to foods by the manufacturer, cook or consumer, plus sugars naturally present in honey, syrups and fruit juices (WHO, 2015). As for the general population, persons with T2D should limit salt intake to 5 g/day.

3.1.1.2 Physical Activity

It has been shown that physical activity improves blood glucose control, reduces cardiovascular risk factors and contributes to weight loss and well-being in people with T2D (ADA, 2021). Exercise is a subset of physical activity that is planned, structured and repeated with the aim of maintaining or enhancing physical fitness (Caspersen et al., 1985). Interventions of at least 8-week duration resorting to structured exercise have been shown to lower A1C by an average of 0.66% in persons with type 2 diabetes, even without a significant change in body mass index (BMI) (ADA, 2021). Particularly in older adults with type 2 diabetes, A1C

Table 3.2 Diet: low-level target behaviours in T2D

Low-level target behaviours	Rationale	Level of evidence[a]
Reduce calorie intake in persons with overweight or obesity	Reduction of calorie intake in combination with lifestyle modification is expected to reduce weight at least by 5% Weight loss of at least 5% produces beneficial outcomes in glycaemic control, lipids and blood pressure	A
Eat nutrient-dense carbohydrate sources that are high in fibre (at least 14 g fibre per 1000 kcal) and minimally processed[b]	Adequate carbohydrate intake, namely, reducing overall intake when excessive, has demonstrated to improve glycaemia in persons with diabetes Eating plans should emphasise non-starchy vegetables[c], fruits and whole grains, as well as dairy products, with minimal added sugars	B
Minimise the consumption of foods with added sugar	Foods with added sugar can displace healthier, more nutrient-dense food choices	A
Eat food rich in monounsaturated and polyunsaturated fats[d]	A Mediterranean-style diet rich in monounsaturated and polyunsaturated fats may improve glucose metabolism and lower cardiovascular disease risk; it can be an effective alternative to a low-fat diet	B
For those who drink alcohol, drink with moderation[e]	Alcohol consumption may place people with diabetes at increased risk for hypoglycaemia, especially if taking insulin or insulin secretagogues People with diabetes should know about the signs, symptoms and self-management of delayed hypoglycaemia after drinking alcohol	B B
Prefer water in detriment of sugar-sweetened beverages or non-nutritive-sweetened beverages	Avoiding sugar-sweetened beverages, including fruit juices, is associated with glycaemic and weight control and reduction of the risk of cardiovascular disease and fatty liver Low-calorie, non-nutritive-sweetened beverages – assuming non-nutritive sweeteners' role in reducing overall energy and carbohydrate intake – may be used as an occasional or short-term option Water intake and decreasing both sweetened and non-nutritive-sweetened beverages are encouraged	B

Based on ADA (2021)

[a]For details about the evidence levels, refer to https://professional.diabetes.org/content-page/practice-guidelines-resources

[b]Examples for high-fibre, nutrient-dense carbohydrate sources are fruits, minimally processed cereals and cereal products (e.g. bread) and legumes (e.g. beans, chickpeas, lentils, peas, broad beans)

[c]In fact, all vegetables have some starch or other forms of carbohydrates. Corn, potatoes, cassava, yams and green peas are examples of vegetables with higher carbohydrate content, while lettuce, spinach, cabbage and watercress have lower carbohydrate content

[d]Vegetable oils, pulses, seeds and nuts are sources for both mono- and polyunsaturated fatty acids

[e]Drinking alcohol in moderation can be considered as no more than one drink per day for adult women and no more than two drinks per day for adult men. One drink corresponds roughly to 355 ml of beer, 150 ml of wine and 36 ml of distilled spirits

levels decrease with resistance training (ADA, 2021), regarded as exercise to stimulate the muscles and enhance strength resorting to external resistances or one's own body weight.

Globally, daily physical activity is recommended to decrease insulin resistance. Over time, activities should progress in intensity, frequency and/or duration to at least 150 minutes/week of moderate-intensity physical activity (ADA, 2021). A goal of 7000 steps/day is consistent with recommendations of 150 minutes/week of moderate-to-vigorous physical activity (Tudor-Locke et al., 2011b); specifically, this step-count threshold has shown benefit in persons with type 2 diabetes (Rossen et al., 2021). Younger and more physically fit persons able to run at 9.7 km/h for at least 25 min can benefit from shorter-duration, higher-intensity, activity (75 minutes/week). Aerobic physical activity[2] bouts should ideally last at least 10 minutes, with the goal of 30 minutes per day or more, most days of the week (ADA, 2021).

Physical activity has conditional recommendations for persons with cardiovascular risk factors. These risk factors, as well as atypical presentation of coronary artery disease, should be previously assessed in persons with diabetes. High-risk persons should be encouraged to start physical activity with short periods of low-intensity exercise and slowly increase the intensity and duration as tolerated (ADA, 2021). Lower-level target behaviours in physical activity are described in Table 3.3.

3.1.1.3 Smoking Cessation

Smoking cessation means quitting tobacco use or e-cigarettes. For people with T2D, the risk of cardiovascular disease, premature death, microvascular complications and worse glycaemic control is significantly increased by smoking, as shown by evidence from generalisable randomised controlled trials (level of evidence A) and supportive evidence from cohort studies (level of evidence B) (ADA, 2021). Therefore, smoking cessation is strongly advised for this population.

3.1.1.4 Medication Adherence

Medication adherence is an important determinant of outcomes in persons with diabetes, by decreasing morbidity and mortality. This includes adherence to antidiabetic agents and, where necessary, antihypertensive agents, statins, antiplatelet agents, nicotine replacement therapy and other medicines (ADA, 2021).

The rationale for taking specific drugs should be discussed with the person on a per case basis.

---

[2]Aerobic physical activity is any bodily movement produced by skeletal muscles that results in energy expenditure (Caspersen et al., 1985) mainly provided by oxidation of lipids and carbohydrates. This includes all sorts of daily living, worksite, fitness and sport-related physical activities (e.g. walking, jogging, climbing stairs and many other).

Table 3.3 Physical activity: low-level target behaviours in physical activity in T2D

Low-level target behaviours	Rationale	Level of evidence[a]
Engage in 150 min or more of moderate-to-vigorous intensity aerobic activity per week, spread over at least 3 days/week, with no more than 2 consecutive days without activity *OR for younger and more physically fit individuals*: engage in a minimum 75 min/week of vigorous-intensity or interval training	Moderate to high volumes of aerobic activity are associated with substantially lower cardiovascular and overall mortality risks Daily exercise, or at least not allowing more than 2 days to elapse between exercise sessions, decreases insulin resistance	B
Engage in 2–3 sessions/week of resistance exercise on non-consecutive days	Resistance training of any intensity may improve glycaemic control and strength	B
Decrease the amount of time spent in daily sedentary behaviour	Avoiding extended sedentary periods may help in glycaemic control	B
Interrupt prolonged sitting every 30 min	Prolonged sitting should be interrupted every 30 minutes for glycaemic control benefits	C
For older adults with diabetes, train flexibility and balance 2–3 times/week	A wide range of activities, including yoga, tai chi and other types, can have significant impacts on A1C, flexibility, muscle strength and balance	C

Based on ADA (2021)
[a]For details about the evidence levels, refer to https://professional.diabetes.org/content-page/practice-guidelines-resources

3.1.1.5 Symptom Monitoring and Management

Hypoglycaemia is the most frequent and life-threatening adverse effect of medicines such as insulin or insulin secretagogues. In type 2 diabetes, it is potentially preventable (Silbert et al., 2018). Persons with type 2 diabetes are expected to identify symptoms suggestive of hypoglycaemia, i.e. monitoring and managing them, by taking the appropriate course of action.

Hypoglycaemia causes neurogenic and neuroglycopenic symptoms. The neurogenic symptoms result from the activation of the autonomic nervous system in response to a reduction in blood glucose; these symptoms include (Cryer, 2019):

- Tremor, palpitations and anxiety/arousal (catecholamine-mediated, adrenergic).
- Sweating, hunger and paresthesias (acetylcholine-mediated, cholinergic).

The neuroglycopenic symptoms result from the brain's deprivation of glucose; they include dizziness, weakness, drowsiness, delirium, confusion and, for lower blood sugar levels, seizure and coma.

Neuroglycopenic symptoms may be more difficult to perceive. Typically, neurogenic symptoms develop at a higher blood glucose threshold (approximately 58 mg/dL or 3.2 mmol/L), while neuroglycopenic symptoms develop at lower blood glucose threshold (approximately 51 mg/dL or 2.7 mmol/L), respectively, although there is intra- and inter-individual variation (Tesfaye et al., 2010).

A rapidly available glucose source should be chosen (e.g., table sugar) for correcting hypoglycaemia, as both protein and fat can influence the absorption rate for glucose (ADA, 2021). The self-management of hypoglycaemia in persons with blood glucose <70 mg/dL [3.9 mmol/L] relies preferably on the intake of glucose (15–20 g), although any form of carbohydrate that contains glucose may be used. Fifteen minutes after glucose intake, self-monitoring of blood glucose, where available, should be performed; if hypoglycaemia is maintained, glucose consumption (15 g) should be repeated. This approach is known as the "rule of 15". Once blood glucose returns to normal, the person should eat a meal or snack to prevent recurrence of hypoglycaemia (ADA, 2021).

3.1.2 Chronic Obstructive Pulmonary Disease

Chronic obstructive pulmonary disease (COPD) is currently one of the top three causes of death worldwide, with 90 per cent of fatalities occurring in low- and middle-income nations (GOLD, 2021). It is a common, preventable and treatable disease characterised by persistent respiratory symptoms and airflow limitation, caused by airway and/or alveolar abnormalities, which are typically originated by significant exposure to noxious particles or gases and influenced by host factors, such as abnormal lung development (GOLD, 2021).

Self-management of COPD, including the use of medicines, lifestyle changes and avoidance of risk factors, is essential to prevent disease progression and improve health outcomes. Evidence-based self-management principles are recommended in clinical guidelines for COPD (GOLD, 2021).

3.1.2.1 Physical Activity

The 2022 report of the Global Initiative for Chronic Obstructive Lung Disease (GOLD, 2021) recommends physical activity for persons with COPD and evidence of benefits are considered clear. However, promoting and maintaining physical activity is challenging, and these persons have generally reduced physical activity levels, which increases the risk of adverse outcomes, including hospitalisation rates and mortality (GOLD, 2021). One barrier for broader physical activity promotion in persons with COPD is linked with the lack of evidence-based guidelines and inconsistencies regarding the type, quantity, timing and other cornerstone components of physical activity recommendations (GOLD, 2021; Liguori, 2021).

Assessment of persons with COPD has been focused mainly on exercise tolerance (GOLD, 2021) and not so much on physical activity. In this respect, a reference measure is the number of steps performed per day (Tudor-Locke et al., 2011a). A meta-analysis showed that the simple use of a pedometer was enough to increase daily physical activity in persons with COPD, particularly in those with higher physical activity levels ($\geq$4000 steps/day) at baseline, who probably had higher exercise tolerance (Armstrong et al., 2019). This beneficial effect was observed in the absence of pulmonary rehabilitation programmes, which are synergic to the use of pedometers. These findings underscore the importance of exercise tolerance assessment for tailoring lifestyle intervention in these persons and suggest that a stepwise approach may be sensible: focusing first on increasing exercise tolerance of beginners or persons with more limited physical activity and implementing physical activity promotion interventions for those with higher exercise tolerance and physical activity capacity.

3.1.2.2 Smoking Cessation

Smoking cessation is key to reducing progressive decline in lung function over time, as well as exacerbations (level of evidence A[3]) and smoking-related co-morbidities, such as lung cancer and cardiovascular disease (GOLD, 2021). Globally, cigarette smoking is the most common, but other types of tobacco such as pipe or cigar should be also considered.

3.1.2.3 Medication Adherence

Medication adherence is an important determinant of outcomes in persons living with COPD, by reducing symptoms, the frequency and severity of exacerbations, decreasing the risk of hospital admission and improving exercise tolerance and health status (GOLD, 2021). This includes adherence to bronchodilators, antimuscarinic drugs, anti-inflammatory agents, oxygen and, where necessary, other medicines (GOLD, 2021).

Adherence to influenza and pneumococcal vaccination can reduce the incidence of lower respiratory tract infections and death in persons with COPD (level of evidence B[4]) (GOLD, 2021).

The rationale for adhering to specific drugs should be discussed with the person on a per case basis.

3.1.2.4 Symptom Monitoring and Management

Persistent respiratory symptoms in COPD include dyspnoea, cough and/or sputum production (GOLD, 2021). There are recommendations to self-monitor and manage both dyspnoea and exacerbations, based on written action plans (GOLD, 2021).

---

[3] For details about the evidence levels, refer to https://goldcopd.org/gold-reports/

[4] For details about the evidence levels, refer to https://goldcopd.org/gold-reports/

In monitoring dyspnoea, trends and changes in symptoms such as respiratory distress, tiredness, restriction of activity and sleep disturbance are more meaningful than specific measures (GOLD, 2021). Integrated self-management of dyspnoea, based on a written action plan, may include breathlessness and energy conservation techniques and stress management strategies.

Exacerbation is an acute worsening of respiratory symptoms, usually associated with increased airway inflammation, increased mucus production and marked gas trapping. Dyspnoea, increased sputum purulence and volume, cough and wheeze are common symptoms of exacerbations. Written action plans for exacerbations must be personalised with respect to how to avoid aggravating factors and how to monitor and manage worsening of symptoms and contact information in the event of an exacerbation (GOLD, 2021).

3.1.3 Arterial Hypertension

Arterial hypertension is characterised by extra strain on blood vessels, heart and other organs. Persistent high blood pressure can increase the risk of several serious and potentially life-threatening conditions, such as myocardial infarction, haemorrhagic stroke and heart failure.

Changes in lifestyle may be sufficient to delay or prevent the need for drug therapy in persons with grade 1 hypertension and can also augment the effects of antihypertensive agents (Williams et al., 2018). Lifestyle changes (e.g. diet, physical activity, smoking cessation) are therefore considered in reference guidelines as the standard and primary treatment for controlling hypertension (Whelton et al., 2018; Williams et al., 2018). The American College of Cardiology, the American Heart Association, and others also additionally recommended that newly diagnosed hypertension with a lower blood pressure threshold can be managed with lifestyle antihypertensive therapy rather than medicines (Whelton et al., 2018). Difficulties in maintaining these lifestyle changes over time are a major drawback of this approach (Williams et al., 2018).

The following sections describe the lower-level target behaviours relevant for the self-management of arterial hypertension, based on Williams et al. (2018).

3.1.3.1 Diet

Diets mainly composed of vegetables, fruits, whole grains, low-fat dairy products, fish, chicken and lean meats, which are high in minerals, fibre and unsaturated fatty acids (namely, monounsaturated from olive oil) and low in saturated fat and cholesterol, are associated with a reduction of blood pressure (Williams et al., 2018). One example is the Mediterranean diet and Dietary Approaches to Stop Hypertension (DASH).

Examples of diet lower-level target behaviours are described in Table 3.4.

Table 3.4 Diet: low-level target behaviours in hypertension

Low-level target behaviours	Rationale	Level of evidence[a]
Limit salt intake to <5 g per day	Excess sodium intake (>2 g of sodium per day or >5 g of salt daily) has shown an increase in blood pressure with age	A
Increase the consumption of fresh fruits and vegetables	Adopting a healthy and balanced diet may assist in blood pressure reduction and reduce cardiovascular risk	A
Increase the consumption of fish		
Prefer monounsaturated and polyunsaturated fat food sources[b]		
Minimise the consumption of red meat, preferring fish and poultry		
For those who drink alcohol, drink with moderation[c]	Moderate consumption of alcohol is associated with a reduction in cardiovascular events and all-cause mortality	A

Based on Williams et al. (2018)

[a]For details about the evidence levels, refer to https://doi.org/10.1080/08037051.2018.1527177
[b]Vegetable oils, pulses, seeds and nuts are sources for both mono- and polyunsaturated fatty acids
[c]Another way of defining moderate alcohol consumption is drinking alcohol less than 14 units per week for men and less than 8 units per week for women. A unit of alcohol is equal to 125 ml of wine or 250 ml of beer (Tasnim et al., 2020)

3.1.3.2 Physical Activity

Aerobic physical activity has been considered particularly beneficial for both the prevention and treatment of hypertension and to lower cardiovascular risk and mortality (Williams et al., 2018). Recommendations for physical activity in persons with hypertension (Table 3.5) are similar to those for the general population, except for vigorous exercise,[5] which is not endorsed (Whelton et al., 2018; Williams et al., 2018). Resistance exercise[6] is also recommended for this population (Williams et al., 2018).

---

[5]Vigorous exercise refers to exercise that is performed at 6.0 or more METs (1 MET ≈ resting metabolism), meaning it demands 6 times or more the energy expenditure and oxygen consumption, compared to that of one's resting state (WHO, 2020). On an individual's personal capacity perspective, vigorous-intensity exercise is usually a 7 or 8 on a scale of 1 to 10 (WHO, 2020). In a practice perspective, vigorous exercise includes but is not limited to activities like jogging or running, playing strenuous racquet sports (singles tennis, paddle ball or others) or other strenuous sports (basketball, soccer or others) (Sallis et al., 1985).

[6]Resistance exercise is any exercise performed to optimise muscular fitness, namely, strength, hypertrophy, power and local muscular endurance, which can encompass free weights, machines, body weight, bands/tubing or any other object that requires one to exert force against a resistance (Ratamess, 2021).

Table 3.5 Physical activity: low-level target behaviours in hypertension

Low-level target behaviours	Rationale	Level of evidence[a]
Performance at least 30 min of moderate-intensity dynamic aerobic exercise (walking, jogging, cycling or swimming) on 5–7 days per week	Aerobic endurance training, dynamic resistance training and isometric training[b] reduce resting systolic blood pressure and diastolic blood pressure	A
Performance of resistance exercises on 2–3 days per week	Resistance exercises on 2–3 days per week reduce blood pressure in persons with hypertension	A

Based on Williams et al. (2018)
[a]For details about the evidence levels, refer to https://doi.org/10.1080/08037051.2018.1527177
[b]Isometric training is all training performed with voluntary muscle contraction but without movement (i.e. sustained contraction against an immovable load or resistance with no or minimal change in length of the involved muscle group) (Fu et al., 2020)

3.1.3.3   Smoking Cessation

Persons with normal blood pressure and untreated hypertension that smoke present higher daily blood pressure values than non-smokers (Williams et al., 2018). Smoking cessation is the most cost-effective cardiovascular disease prevention strategy, including for the prevention of stroke, myocardial infarction and peripheral artery disease (Williams et al., 2018).

3.1.3.4   Medication Adherence

Medication adherence in persons with hypertension is associated with an effective reduction of blood pressure and cardiovascular events (Williams et al., 2018). Antihypertensive agents include angiotensin-converting enzyme inhibitors, angiotensin receptor blockers, calcium channel blockers and diuretics (Williams et al., 2018).

The rationale for adhering to specific drugs should be discussed with the person on a per case basis.

3.1.4   Heart Failure

Heart failure is characterised by key symptoms (e.g. dyspnoea, ankle swelling and weariness) that may be accompanied with signs (e.g. elevated jugular venous pressure, pulmonary crackles and peripheral oedema). Elevated intracardiac pressures and/or insufficient cardiac output at rest and/or during exercise is caused by structural and/or functional abnormalities of the heart. Heart failure is one of the most frequent causes of hospitalisation and death. To avoid exacerbations, several self-management behaviours are recommended.

3.1.4.1 Diet

There are three low-level dietary behaviours recommended for persons with heart failure, additionally to a healthy diet: avoiding excessive salt intake (>5 g/day), avoiding large volumes of fluid intake and avoiding excessive alcohol intake, understood as drinking more than 2 units of alcohol per day in men or 1 unit of alcohol per day in women. Evidence shows that limiting sodium intake prevents morbidities associated with this condition (McDonagh et al., 2021).

3.1.4.2 Physical Activity

Persons with heart failure should undertake regular exercise and be physically active (McDonagh et al., 2021). However, specific physical activity guidance for these persons is not available; general physical activity guidelines apply, which recommend ≥150–300 min/week of moderate-intensity aerobic physical activity or 75–150 min/week of vigorous-intensity aerobic physical activity (Piercy et al., 2018; Ye et al., 2021). Persons living with heart failure report a lower involvement in physical activity, which increases their cardiometabolic risk (Yates et al., 2017; Ye et al., 2021).

Exercises in the form of cardiac rehabilitation programmes specific for this population have demonstrated benefit but are outside the remit of this book (McDonagh et al., 2021).

3.1.4.3 Smoking Cessation

Smoking cessation is recommended for persons with heart failure. Although smoking cessation has not been shown to reduce the risk of developing heart failure, tobacco is associated with the development of cardiovascular disease (McDonagh et al., 2021; Ezekowitz et al., 2017).

3.1.4.4 Medication Adherence

Medication adherence is critical to prevent or delay the development of overt heart failure, manage symptomatic persons with reduced ejection fraction and improve exercise capacity. Medication adherence reduces the risk of heart failure-related hospitalisation in persons with signs and/or symptoms of congestion, reduces mortality and prolongs life. This includes adherence to diuretics, antihypertensives, beta-blocker agents and, where necessary, other medicines (McDonagh et al., 2021).

The rationale for adhering to specific drugs should be discussed with the person on a per case basis.

3.1.4.5 Symptom Monitoring and Management

Monitoring and managing symptoms in persons with heart failure are recommended as part of an overall strategy to reduce the risk of hospitalisation and mortality associated with this disease (McDonagh et al., 2021). For example, in the case of increasing dyspnoea or oedema in three consecutive days, persons may increase their diuretic dose and/or alert their healthcare team. Symptom management may include flexible use of diuretic and fluid intake as agreed with the healthcare team. It is also recommended that a sudden unexpected weight gain of >2 kg in three consecutive days triggers an increase in diuretic dose and/or an alert to the healthcare team (McDonagh et al., 2021).

3.1.5   Obesity

Obesity is a metabolic chronic disease resulting from complex interactions between biological, behavioural, psychosocial, genetic and environmental factors, which should be considered by professionals to support self-management (Frühbeck et al., 2019). Understanding the role of these factors is critical to avoid misconceptions, such as associating obesity with a poor lifestyle and absence of will power, and to avoid stigma.

The self-management of obesity includes a diet with restriction of energy intake, regular physical activity and, when applicable, medication-taking. Behaviour changes are by far the most important component for integrating successful eating and activity patterns over the long term (Yumuk et al., 2015). A personalised energy-restricted diet, based on nutritional habits, physical activity, co-morbidities and previous dieting attempts, is beyond the remit of this book, as it requires the intervention of a nutritionist. As for the remainder sections, we focus on recommendations that can be actioned by a range of professionals.

Weight management is a lifelong process; monitoring the weight is important in self-management for persons with obesity. A 5–15% weight loss over a period of 6 months is realistic and of proven health benefit, and a greater (20% or more) weight loss may be considered for those with greater degrees of obesity (BMI $\geq$ 35 kg/m^2) (Yumuk et al., 2015).

3.1.5.1 Diet

Balanced hypocaloric diet produces clinically significant weight loss, irrespective
of the macronutrients' proportion (Yumuk et al., 2015). Low-level target behaviours
depicted in Table 3.6 are underpinned by less robust evidence (in the case, non-
analytic studies, such as case reports and case series, and expert opinion); in the
absence of more robust evidence, they are accepted as desirable for setting self-
management recommendations for adults with obesity.

Decreasing energy density of foods and drinks may encompass a range of
behaviours:

- Adequate the consumption of vegetables, beans, legumes, lentils, grain, unsweet-
  ened cereals, fruits and fibre.
- Prefer the consumption of seafood over meats and meat products with a high-fat
  content.
- Reduce the intake of foods containing added sugars and fats (e.g. spreadable fats,
  fats for seasoning).
- Replace the consumption of sugary drinks and alcohol-containing beverages for
  water or herbal teas.

Additional behaviours advocated in the literature are resorting to a Mediterranean
"type" diet, based on low-fat ingredients, controlling the fat in cooking (sautéed,
stewed or roasted dishes), using aromatic herbs instead of salt, and choosing whole
grain foods (Esposito et al., 2011).

Adopting regular meal patterns and planning what to eat during the day is an
important low-level target behaviour. Persons with obesity benefit from maintaining
a pattern of at least three meals, breakfast, lunch and dinner, as anchor meals
(Camolas et al., 2015). Adopting intermediate snacks between meals may be an
option if effective in appetite control (Chapelot, 2011; Camolas et al., 2015).

Table 3.6 Diet: low-level target behaviours in obesity

Low-level target behaviours	Rationale	Level of evidence[a]
Decrease energy density of foods and drinks	Weight loss between 5% and 15% of initial body weight is associated with a reduction in cardiovascular risk factors, improvement in lipid profiles, reduction in blood glucose and glycosylated haemoglobin and decreased risk for developing type 2 diabetes and other obesity-related complications	Levels 3 and 4
Decrease the size of food portions		
Adopt a regular meal pattern		

Based on Yumuk et al. (2015)

[a]For details about the evidence levels refer to https://doi.org/10.1159/000442721

3.1.5.2 Physical Activity

Increased physical activity, including exercise, is considered a main component of obesity management, in conjunction with a decreased energy intake and healthy eating (Table 3.7). Comprehensive lifestyle interventions are considered effective for weight reduction but also for preserving fat-free mass (Yumuk et al., 2015). Physical activity has proven useful for the management and treatment of most obesity co-morbidities, as well as other diseases and impairments (Pedersen & Saltin, 2015). Furthermore, obesity management may reduce the need to treat co-morbidities pharmacologically (Yumuk et al., 2015).

Guidelines recommend that persons with obesity should engage in at least 150 min/week of moderate aerobic exercise (such as brisk walking), combined with two/three weekly sessions of resistance exercise, to increase muscle strength (Schutz et al., 2019; Yumuk et al., 2015). More physical activity (e.g. about 300 min/week of endurance activity at moderate intensity or 150 min of vigorous activity) may provide additional benefits, including higher mobilisation of visceral fat (Schutz et al., 2019). Additionally, guidelines underscore the importance of enjoyment in reducing sedentarism and increasing physical activity (Schutz et al., 2019). Physical activity can be split into multiple short segments of 10 minutes minimum, to have a metabolic impact (Schutz et al., 2019).

Examples of lower-level target behaviours are described in Table 3.8, based on Yumuk et al. (2015).

Persons with obesity who reach a high cardiorespiratory fitness have a lower mortality risk due to all aetiologies than normal-weight sedentary persons, namely, those who sit or lie down for a prolonged period (Schutz et al., 2019). Walking is suggested as the best physical activity for these persons, as it fosters autonomy and competence and is practical (can be done anywhere, free of charge, without requiring specific equipment, apart from regular walking shoes). Furthermore, persons can adjust the intensity (speed, slopes, stairs) and select a particular terrain (snow, sand, grass or others) (Schutz et al., 2019).

Table 3.7 Approach to the initial management of obesity

| BMI (kg/m²)[a] | WC (cm)[a] | | With co-morbidities |
	Normal	High	
25.0–29.9	L	L	L ± D
30.0–34.9	L	L ± D	L ± D ± S[b]
35.0–39.9	L ± D	L ± D	L ± D ± S
≥40.0	L ± D ± S	L ± D ± S	L ± D ± S

Note: *BMI* body mass index, *WC* waist circumference, *L* lifestyle intervention (diet and physical activity), *D* consider pharmacological treatment, *S* consider bariatric surgery
[a]BMI and WC cut-offs are different for some ethnic groups
[b]Patients with type 2 diabetes on an individual basis. Based on Yumuk et al. (2015)

Table 3.8 Physical activity: low-level target behaviours in obesity

Low-level target behaviours	Rationale	Class of recommendation[a]
Performance at least 150 min/week of aerobic exercise combined with 2–3 weekly sessions of resistance training	Aerobic training (e.g. brisk walking) is the optimal mode of exercise for reducing fat mass and body mass, while resistance training increases lean mass in middle-aged and overweight/obese individuals	{Level 2; grade B}
	Increasing physical activity induces: Reductions in intra-abdominal fat and increases in lean (muscle and bone) mass while attenuating the weight loss-induced decline of resting energy expenditure[b] {level 2}. Reductions in blood pressure and improvements in glucose tolerance, insulin sensitivity, lipid profile and physical fitness {level 1}. Better compliance to the dietary regimen, which has a positive influence on long-term weight maintenance {level 2}. Feelings of well-being and better self-esteem {level 2}. Reductions in anxiety and depression {level 2}.	{Level 1, 2}

Based on Yumuk et al. (2015)

[a]For details about the evidence levels, refer to https://doi.org/10.1159/000442721

[b]Weight loss-induced decline of resting energy expenditure is a decline in resting and exercise-related energy expenditure observed in response to weight loss and negative energy balance (Dulloo et al., 2012). This reduction is greater than the expected reduction in energy expenditure due to the reduction of overall body mass, including both fat and fat-free mass. This thermogenic adaptation (modulated energy expenditure) is believed to be mediated mainly by the autonomic nervous system in response to both reduction of body mass and reduction of energy intake (Dulloo et al., 2012)

3.1.5.3   Medication Adherence

Obesity management requires a holistic approach. Adequate diet and physical activity are the cornerstone of obesity management; pharmacologic therapy has the potential to enhance results obtained by lifestyle interventions (Yumuk et al., 2015). All medicines have inherently more risks than diet and physical activity.

The rationale for adhering to specific drugs should be discussed with the person on a per case basis.

3.1.6   Asthma

Asthma is a complex condition marked by persistent inflammation of the airways. Clinically, it is characterised by respiratory symptoms, such as wheezing, shortness of breath, chest tightness and cough, with varying intensity and frequency over time, as well as fluctuating expiratory airflow restriction. Globally, asthma is a prevalent chronic condition, affecting 1–18% of the population (GINA, 2021).

Diet, smoking cessation, medication adherence and symptom monitoring and management are self-management behaviours that impact on disease management. Regular and moderate physical activity has significant health benefits for people with asthma, as for the general population (GINA, 2021), but it is not regarded as a target behaviour for self-management.

3.1.6.1   Diet

A healthy diet is recommended for persons living with asthma, including high consumption of fresh fruit and vegetables.[7] In some studies, this is linked to an improvement in asthma control and a reduced risk of exacerbations (GINA, 2021). For confirmed food allergy, food allergen avoidance may reduce asthma exacerbations (level of evidence D[8]) (GINA, 2021).

3.1.6.2   Smoking Cessation

Smoking cessation, including e-cigarettes, is a key target behaviour in people with asthma. These persons should be referred to counselling and smoking cessation programmes (level of evidence A) (GINA, 2021). Those with asthma exposed to tobacco have a greater risk of exacerbations when compared with those with asthma that do not smoke (GINA, 2021). Additionally, persons living with asthma should avoid environmental smoke exposure (level of evidence B[9]).

---

[7] The World Health Organization (2003) proposes the consumption of 400 g of fruits and vegetables as an adequate goal for a prudent diet. Two to three servings of fruit (e.g. one pear, one apple, one slice of pineapple) and three servings of vegetables (e.g. soup, salad, boiled vegetables) may be adequate to achieve the goal.

[8] For details about the evidence levels, refer to https://ginasthma.org/wp-content/uploads/2021/05/GINA-Main-Report-2021-V2-WMS.pdf

[9] For details about the evidence levels, refer to https://ginasthma.org/wp-content/uploads/2021/05/GINA-Main-Report-2021-V2-WMS.pdf

3.1.6.3 Medication Adherence

Adherence to medication is central to achieve optimal asthma outcomes by decreasing the frequency of exacerbations (GINA, 2021). This includes adherence to asthma medicines, such as inhalers, and vaccines (e.g. influenza and pneumococcal). In addition, being adherent to a correct inhaler technique is a key low-level behaviour for improving asthma control (level of evidence A[10]).

The rationale for adhering to specific drugs should be discussed with the person on a per case basis.

3.1.6.4 Symptom Monitoring and Management

Asthma symptoms such as wheezing, tightness of the chest, shortness of breath and coughing vary in intensity and frequency and contribute to the disease burden. Insufficient control of symptoms is strongly linked to an increase of exacerbations (GINA, 2021). Monitoring and self-managing symptoms is therefore a key target behaviour (GINA, 2021).

Persons with asthma should be trained to keep track of their symptoms, with or without a diary, and to take action in case of worsening (GINA, 2021). Based on written action plans, agreed with a healthcare professional, these persons can make short-term changes to their treatment for symptom managing.

Self-management of asthma symptoms may include breathing exercises as a supplement to asthma pharmacotherapy; these are beneficial for symptoms and quality of life (level of evidence A), but do not improve lung function or reduce exacerbation risk (GINA, 2021). Relaxation strategies may also be useful to self-manage asthma symptoms (level of evidence B), in conjunction with pharmacotherapy, although there is lack of evidence on the best stress reduction strategy (GINA, 2021).

Preventing asthma symptoms from worsening may include avoidance of allergen exposure (level of evidence C), of outdoor air pollutants and of weather conditions (level of evidence D). Staying indoors in a climate-controlled environment and refraining from vigorous outdoor physical activity may be useful when the weather is very cold or there is significant air pollution. Keeping away from polluted environments during viral infections is also recommended, if feasible (GINA, 2021).

---

[10] For details about the evidence levels, refer to https://ginasthma.org/wp-content/uploads/2021/05/GINA-Main-Report-2021-V2-WMS.pdf

3.1.7   Ischaemic Heart Disease

Ischaemic heart disease, also known as coronary heart disease or coronary artery disease, is a condition marked by the formation of obstructive or nonobstructive atherosclerotic plaque in the epicardial arteries. The dynamic nature of the coronary artery disease process causes a variety of clinical manifestations, which can be classified as acute coronary syndromes or chronic coronary syndromes. This section focuses only on self-management behaviours for chronic coronary syndrome. This syndrome can be sped up or slowed down by lifestyle changes, pharmacologic therapy and invasive procedures. A healthy lifestyle improves the cardiometabolic profile and reduces the risk of cardiovascular events and mortality (Knuuti et al., 2020). Key self-management behaviours include diet, physical activity, smoking cessation, medication adherence and symptom monitoring and management (Knuuti et al., 2020).

3.1.7.1   Diet

Unhealthy diets are the main cause of inception and progression of coronary artery disease. Persons with ischaemic heart disease should adopt a healthy diet and limit alcohol intake (Knuuti et al., 2020). Adequate fat intake (total, saturated and unsaturated) should be achieved by individually adequate portions of lean meat, low-fat dairy products and vegetable oils.

Table 3.9 presents low-level diet behaviours for these persons based on Knuuti et al. (2020).

3.1.7.2   Physical Activity

Physical activity, including exercise, has been advocated for persons with ischaemic heart disease due to beneficial effects on cardiovascular risk factors and cardiovascular system physiology (Knuuti et al., 2020). Increasing exercise capacity is an independent predictor of increased survival among men and women with ischaemic heart disease (Knuuti et al., 2020). Although physical activity is generally considered to be safe, guidelines have stressed the importance of initial assessment for proper physical activity counselling and exercise prescription (Piopoli et al., 2014).

It is recommended that all persons with ischaemic heart disease engage in 30–60 minutes of moderate physical activity most days, but even irregular activity is beneficial (level of evidence A[11]) (Knuuti et al., 2020).

---

[11] For details about the evidence levels, refer to https://doi.org/10.1093/eurheartj/ehz425

Table 3.9 Diet: low-level target behaviours in ischaemic heart disease

Low-level target behaviours	Rationale
Increase consumption of fruits and vegetables (>200 g each per day)[a]	Changes to healthy eating patterns in people with chronic coronary syndromes have resulted in a decrease in mortality and cardiovascular events
Eat 35–45 g of fibre per day[b], preferably from whole grains	
Eat 1–2 servings of fish per week (one to be oily fish)	
Limit saturated fat to <10% of total intake[c]	
Limit consumption of lean meat, low-fat dairy products and liquid vegetable oils	
Adopt a diet rich in monounsaturated and polyunsaturated fats[d]	
As little intake of trans unsaturated fats[e] as possible, preferably no intake from processed food	
Limit salt consumption to 5 g/day	
Avoid energy-dense foods, such as sugar-sweetened soft drinks	
Limit alcohol to <100 g/week or 15 g/day[f]	Levels of alcohol >100 g per week were linked to an increased risk of all-cause and other cardiovascular disease death

Based on Knuuti et al. (2020)

[a]Two to three servings of fruit (e.g. one pear, one apple, one slice of pineapple) and three servings of vegetables (e.g. soup, salad, boiled vegetables) may be adequate to achieve the goal

[b]In addition to preferring wholegrain cereal products, an adequate consumption of fruits and vegetables and legumes would contribute to achieving the daily fibre goals

[c]The only direct way to assess this is to measure the entire food intake and its composition. A practical way to achieve it is to reduce fatty meat, butter, palm and coconut oil, cream, cheese, ghee and lard

[d]Unsaturated fats are found in fish, nuts and seeds, avocado and vegetable oils (e.g. sunflower, soybean, canola and olive oils)

[e]Dietary trans-fats can be derived either from industrially produced foods (e.g. from baked and fried foods, pre-packaged pizzas, pies, cookies, biscuits, wafers and some cooking oils and spreads) or ruminant (e.g. cows, sheep, goats and camels) meat and dairy

[f]Another way of defining alcohol consumption is using weight per unit of of volume. Fifteen grams per day is equivalent to roughly 375 ml of 5% alcohol by volume (ABV) beer, 150 ml of 12.5% ABV wine and 50 ml of 40% ABV spirit

3.1.7.3 Smoking Cessation

Smoking cessation improves the prognosis of chronic coronary syndrome, with a 36% mortality risk reduction for those who quit (Critchley & Capewell, 2003; Hussain et al., 2021). People with ischaemic heart disease should stop smoking and avoid second-hand smoke (Knuuti et al., 2020).

3.1.7.4   Medication Adherence

Drug therapy aims to reduce angina symptoms and ischaemia caused by exercise and to prevent cardiovascular events (Knuuti et al., 2020). Adherence to anti-ischaemic drugs and, where necessary, other drugs, is an important determinant of clinical outcomes, by decreasing morbidity and mortality.

3.1.7.5   Symptom Monitoring and Management

Persons with chronic coronary syndrome experience "stable" anginal symptoms. Typical angina has three characteristics: constricting discomfort in the front of the chest or in the neck, jaw, shoulder or arm (short-term pain, typically less than 10 minutes); precipitation by physical exertion; and relief by rest or nitrates within 5 minutes. Symptoms typically appear or become more severe with higher degrees of exertion – such as walking up an incline or against a breeze or in cold weather – and then disappear in a matter of minutes after these triggering variables have passed. 'Stable' anginal symptoms may include shortness of breath and less-specific symptoms, such as fatigue, nausea, burning, restlessness, or a sense of impending doom (Knuuti et al., 2020).

In the self-management of ischaemic heart disease, early recognition of symptoms and avoidance of behaviours that trigger these symptoms are essential to reduce the risk of hospitalisation and mortality, as well as deterioration of associated co-morbidities (Knuuti et al., 2020).

3.2   Assessing Target Behaviours in the Self-Management of Chronic Diseases

Assessment is a first step in understanding self-management behaviours and identifying needs. Available tools can either measure a single target behaviour (e.g. Adult Eating Behaviour Questionnaire, AEBQ) or a range of self-management behaviours. For instance, the Summary of Diabetes Self-Care Activities (SDSCA) is a brief self-report questionnaire of diabetes self-management that includes items assessing diet, physical activity and smoking (Toobert et al., 2000). The European Heart Failure Self-care Behaviour scale (EHFScB-9) is another reliable and valid instrument, with nine items covering physical activity, diet, medication-taking and other behaviours (Jaarsma et al., 2009). Next, we present examples of instruments that are not disease-specific to assess self-management behaviours.

An important consideration for these instruments is that their psychometric properties (i.e. adequate reliability and validity) may not have been evaluated in all countries or settings.

3.2.1 Diet

As highlighted by Hu and Willett (2018), "diet is a complex, dynamic exposure with no perfect method to quantify all aspects of dietary intakes and eating behaviours".

To assess an individual's usual diet in epidemiological studies, Burke (1947) proposed a detailed dietary history interview, which included a 24-hour recall, a 3-day food record and a checklist of foods consumed over the preceding month. To overcome the time- and effort-consuming nature of this method, surrogate methods have been proposed for research contexts. Amidst these are food-frequency questionnaires (Willet, 1990), 24-hour recalls (Buzzard, 1998) and food records (or diaries) (Buzzard, 1998). As any approach relying on self-reporting, data may be biased by memory and willingness to give true reports.

A growing amount of evidence supports the beneficial effect of the quality of diet, over the importance of selected nutrients, in endpoints such as mortality and cardiovascular events (e.g. Panizza et al., 2018). Several instruments have been proposed to estimate the overall diet quality – not necessarily individual foods or nutrients. One example is the Healthy Eating Index (HEI) (Krebs-Smith et al., 2018). The 14-point Mediterranean Diet Adherence Screener (MEDAS) is considered a valid instrument for estimating adherence to the Mediterranean diet (Schröder et al., 2011). Both HEI and MEDAS look at a set of foods, indicative of overall dietary quality, and can be used to assess low-level dietary behaviours.

In clinical settings, 24-hour recalls and food records (or diaries) may be useful for the person's initial assessment and monitoring. Persons may report what they know is appropriate to eat and not their real diet. In research contexts, it has been demonstrated that, e.g. individuals with higher fat and total energy intake are more prone to underreport (Hebert et al., 1995). The influence of social desirability also influences reported intakes of fruit and vegetables (Miller et al., 2008). Nonetheless, these reports represent a surrogate of individuals' diet and provide indication on nutrition knowledge; both are useful as a baseline information in behaviour change interventions.

As summarised in Table 3.10, for assessing dietary behaviours in the context of behaviour change support, professionals may resort to measures of overall diet quality, which have predictive value for health outcomes, such as the HEI and MEDAS. Other options are 24-hour recalls and food records, which may be complemented by objective biomarkers (e.g. blood levels of fatty acids, carotenoids) and digital technology tools (e.g. cell phone images of foods and meals), for improving the accuracy of dietary assessment and the intervention plan.

Table 3.10 Examples of instruments for assessing diet and dietary behaviours in persons living with chronic disease

Food-frequency questionnaires (Willet, 1990)
Food-frequency questionnaires consist of a list of foods and asking subjects to report how often they have eaten each of them (from never to several times a day), over a specific period (e.g. the last 6 months). The accuracy of the data depends on the adequacy of the list (both the type and number of foods listed) and on the memory and cooperation from the person
24-hour recall (Witschi, 1990)
Twenty-four-hour recall is a method that tries to identify and quantify a person's food intake, during a specific day. Its reliability depends on the skill of the interviewer and the memory, cooperation and communication ability of the person
Food records (Buzzard, 1998)
Food record (or diary) is a method for which persons are requested to record everything they eat, using food weight or common measures, such as tablespoons and units, and when they eat it. Accuracy does not depend on the person's memory, evaluation or frankness, although it may be diminished by changes in food choices or eating behaviours
Healthy Eating Index (Reedy et al., 2018)
The healthy eating index estimates diet quality by measuring its alignment with the dietary guidelines for Americans. It may be used to assess diet quality in population surveillance initiatives and evaluations of food environments and food assistance programmes, as well as for nutrition interventions
Mediterranean Diet Adherence Screener (Schröder et al., 2011)
The Mediterranean diet adherence screener (MEDAS) is a 14-item instrument, developed for rapid estimation of adherence to the Mediterranean diet, useful in clinical practice

3.2.2 Physical Activity

Common methods for measuring physical activity include self-report (e.g. logs, diaries, questionnaires, recalls) and objective measures (e.g. accelerometers/activity monitors, pedometers, heart rate monitors, combined sensors) (Dowd et al., 2018; Rodrigues et al., 2022). In general, objective measures are preferable for assessing physical activity in adults (Dowd et al., 2018) and are often used as criterion methods (Rodrigues et al., 2022). They are also credited with a more accurate estimation of daily physical activity levels in persons with chronic diseases (Alothman et al., 2017). Self-reported measures tend to overestimate physical activity and underestimate sedentary time when compared with accelerometry, including in persons with chronic diseases (Schmidt et al., 2020; Ramirez-Marrero et al., 2014).

Self-reported measures of physical activity, also referred to as subjective measures, are usually assessed using questionnaires or recalls, which may be self-administered or answered as an interview. Self-reported physical activity may also be collected using diaries or logs, which require the person to record all activities carried out throughout the day, usually on a minimum of 2 weekdays and 1 weekend day. General questionnaires are simple and easy to administer; examples can be found in Table 3.11. An extensive list of self-reported physical activity assessment tools and the underscoring of their limitations are available from meta-research

Table 3.11 Common methods for assessing physical activity in persons with chronic diseases

Objective measures

Accelerometer

Accelerometers detect movement/acceleration; they are used to count steps and to assess the intensity of acceleration; to estimate time spent in physical activity of light, moderate and vigorous intensity; and to estimate related energy expenditure. The objective measurement of physical activity using accelerometer has been successfully conducted in persons with different clinical conditions, including type 2 diabetes (Feig et al., 2021; Baier et al., 2021), chronic obstructive pulmonary disease (Morita et al., 2018; Pinto et al., 2020), hypertension (Sousa Junior et al., 2020; Schlenk et al., 2021; Martinez Aguirre-Betolaza et al., 2020), heart failure (Schmidt et al., 2020; Izawa et al., 2014), obesity (Baillot et al., 2020; Bell et al., 2015) and ischaemic heart disease (Chokshi et al., 2018)

Pedometer

A pedometer is a device that counts the number of steps taken. The measurement of step counts using pedometers has been used and found useful in persons with different clinical conditions including, but not limited to, type 2 diabetes (Leischik et al., 2021), hypertension (Sousa Junior et al., 2020) and ischaemic heart disease (Reid et al., 2012)

Combined motion sensors

Combined sensors are wearables that include different data collection sensors, including accelerometer (as earlier described), inclinometer (which is important to warrant accuracy in the assessment of sedentary behaviour) and heart rate sensor which has been found useful to assess the metabolic intensity in different movement behaviours, beyond the related time spent in them, and estimate energy expenditure (Strath et al., 2013)

Self-report measures

International physical activity questionnaire (IPAQ)

It consists of a questionnaire originally developed in the late 1990s by an international consensus group, to be used by adults (18–65 years old) in different formats: short and long form, self-administered or telephone interview form and focusing on the "last 7 days" or on a "usual week" (Craig et al., 2003). The IPAQ has been widely used in both healthy and clinical populations, including persons with type 2 diabetes (Lopes et al., 2021), hypertension (Lopes et al., 2021; Riegel et al., 2019), heart failure (Schmidt et al., 2020; Klompstra et al., 2019) and ischaemic heart disease (Maddison et al., 2015)

Global physical activity questionnaire (GPAQ)

The GPAQ consists of a 16-item questionnaire designed to estimate an individual's level of physical activity in 3 domains, work, transport and leisure time, and the time spent in sedentary behaviours, from a "usual week". It was developed by the WHO to be used both at a local and international scale as a surveillance tool to monitor physical activity levels, as a chronic disease risk factor (WHO, 2005), and it distinguishes itself from IPAQ by assessing each domain separately instead of including all domains together for the calculation of physical activity-related parameters. The GPAQ has been widely used in both healthy and clinical populations, including persons with type 2 diabetes (Alzahrani et al., 2019) and obesity (Marcos-Pardo et al., 2020; Baillot et al., 2020)

Yale physical activity survey (YPAS)

Developed initially for the elderly population, to assess overall and specific physical activity and sitting, including index scores composed of questions on vigorous activity, leisurely walking, moving, sitting and standing (Dipietro et al., 1993). YPAS has been further validated and used in other populations, including in with chronic obstructive pulmonary disease (Mihaltan et al., 2019)

(Rodrigues et al., 2022). Objective measures of physical activity (Table 3.11) are available in daily practice via wearables (e.g. activity monitors, smartwatches) and mobile phone-based sensors.

Such measures of physical activity have been commonly used in persons with chronic disease in different research contexts (Table 3.11) and are widely available in daily practice, via wearables (e.g. activity monitors, smartwatches) and mobile phone-based sensors.

3.2.3 Medication Adherence

Medication adherence measures may include direct or indirect assessment of how the person uses prescribed medicines (Lam & Fresco, 2015; Buono et al., 2017; Anghel et al., 2019; Forbes et al., 2018).

Direct methods such as measurement of drug or metabolite levels in blood or urine, or detection of blood levels of biological markers added to the drug formulation, prove drug administration. However, they are subject to variations in inter-individual metabolism (Buono et al., 2017). Additionally, they are relatively expensive, potentially burdensome (e.g. requiring venipuncture or urine collection) and not available for most medicines, which hinders routine use.

Amidst indirect methods, self-report, through questionnaires, diaries or clinical interview, is the most common approach in practice. Questionnaires are inexpensive and simple to apply, contributing to their popularity. They can be self-administered (online or in a paper format) or through an interviewer, depending on the questionnaire. These questionnaires have generally been validated against other measures, both subjective and objective. Some instruments give additional information about attitudes, behaviour determinants and intentions (Anghel et al., 2019), an aspect that will be explored in Chap. 4. Examples of questionnaires are listed in Table 3.12.

Table 3.12 Questionnaires for assessing medication adherence in persons living with chronic disease

---

Brief Medication Questionnaire (Svarstad et al., 1999)
Explores both patient's medication-taking behaviour and barriers to adherence. The questionnaire has three screens: a five-item regimen screen, which assesses how persons took each of their medications in the past week; a two-item belief screen that appraises drug effect and adverse events; and a two-item recall screen related to remembering difficulties

---

Eight-Item Morisky Medication Adherence Scale (MMAS-8) (Morisky et al., 2008)
An eight-item questionnaire: The first seven items are yes/no answers, and the last item is a five-point Likert response, focusing on medication-taking behaviours, especially related to adherence barriers

---

Medication Adherence Report Scale (MARS-10) (Horne & Weinman, 2002; Cohen et al., 2009)
It assesses medication-taking behaviours and attitudes towards medication through ten questions with a simple scoring to evaluate adherence behaviour, attitude towards medication and general disease control during the past week. Additionally, it examines beliefs and barriers to medication adherence.
The MARS-5 and MARS-6 are variations of this questionnaire, including five and six statements, respectively, with a five-point rating scale

Claims data derived from medicines dispensing in pharmacies is another indirect method to assess adherence. Measures calculated from pharmacy data to assess the number of doses dispensed in a time period include PDC (proportion of days covered) and MPR (medication possession ratio). These measures are still not readily available in clinical practice in many countries.

In summary, none of the available methods can be considered as a gold standard, and the combination of methods is recommended, when feasible. Decisions on the selection of a method to assess adherence should be based on a per case basis, considering aspects such as the accuracy and reliability of data and resources available.

3.2.4 Smoking Cessation

Smoking abstinence is the main outcome of smoking cessation interventions. This can be assessed through biochemical validation or self-report measures (including interviews). The main outcomes assessed are point prevalence abstinence (e.g. past 7 days) and prolonged or continuous abstinence (e.g. past 30 days, 6 months and 12 months). The Russell standard criteria (West et al., 2005) are commonly used to select the best measure for assessing smoking abstinence in the context of smoking cessation interventions, in which biochemically validated abstinence is preferred over self-report and prolonged or continuous abstinence is preferred over point prevalence abstinence; in 2017, a Delphi exercise was conducted to achieve a consensus on assessment criteria for smoking cessation (Pound et al., 2021).

Examples of biochemical measures are:

- Carbon monoxide, detected through, e.g. expired air sample.
- Cotinine (e.g. saliva samples).

As for self-report or interviews, *number of cigarettes smoked in the past 7 days (or 30 days)* is the most common. Other questions include:

- Quit attempts in the past 30 days.
- Average cigarettes smoked per day.
- Duration of smoking at that rate.

When assessing smoking cessation, it is also important to include assessment of vaping.

Key Points
- Physical activity, medication adherence and smoking cessation are high-level target behaviours for self-managing type 2 diabetes, chronic obstructive pulmonary disease, hypertension, heart failure, obesity, asthma and ischaemic heart disease.
- Diet is a high-level target behaviour in the self-management of type 2 diabetes, hypertension, heart failure, obesity, asthma and ischaemic heart disease.

- Symptom monitoring and management is a high-level target behaviour relevant for those living with type 2 diabetes, chronic obstructive pulmonary disease, heart failure, asthma and ischaemic heart disease.
- Diet low-level target behaviours include, for example, reducing calorie intake, minimising the consumption of foods with added sugar, avoiding excessive salt intake, increasing the consumption of fresh fruits and vegetables and limiting saturated fat consumption.
- Decreasing daily sedentary behaviour and performing aerobic activity or resistance exercises are examples of low-level target behaviours in physical activity.
- Medication adherence low-level target behaviours include, for instance, adherence to influenza and pneumococcal vaccination, adherence to a correct inhaler technique or adherence to specific drugs.
- Managing symptoms' low-level target behaviours may include flexible use of diuretic and fluid intake or avoidance of allergen exposure or outdoor air pollutants.
- Assessment, preferably with specific and validated tools, is the first step for understanding self-management behaviours and identifying needs.
- Target behaviours can be assessed using objective (e.g. accelerometers/activity monitors, pedometers) and/or subjective measures (e.g. Eight-Item Morisky Medication Adherence Scale, food-frequency questionnaires).

References

ADA – American Diabetes Association. (2019). *Standards of medical care in diabetes-2019.* https://care.diabetesjournals.org/content/diacare/suppl/2018/12/17/42.Supplement_1.DC1/DC_42_S1_2019_UPDATED.pdf

Alothman, S., Yahya, A., Rucker, J., & Kluding, P. M. (2017). Effectiveness of interventions for promoting objectively measured physical activity of adults with type 2 diabetes: A systematic review. *Journal of Physical Activity & Health, 14,* 408–415. https://doi.org/10.1123/jpah.2016-0528

Alzahrani, A. M., Albakri, S. B. B., Alqutub, T. T., Alghamdi, A. A., & Rio, A. A. (2019). Physical activity level and its barriers among patients with type 2 diabetes mellitus attending primary healthcare centers in Saudi Arabia. *Journal of Family Medicine and Primary Care, 8,* 5. https://doi.org/10.4103/jfmpc.jfmpc_433_19

American College of Cardiology Foundation and the American Heart Association, I. (2018). *Guideline for the prevention, detection, evaluation, and management of high blood pressure in adults.* http://hyper.ahajournals.org

American Diabetes Association. (2021). *Professional practice committee: Standards of medical care in diabetes—2021.* https://doi.org/10.2337/dc21-Sppc.

Anghel, L. A., Farcas, A. M., & Oprean, R. N. (2019). An overview of the common methods used to measure treatment adherence. *Medicine and Pharmacy Reports, 92*(2), 117–122. https://doi.org/10.15386/mpr-1201

Armstrong, M., Winnard, A., Chynkiamis, N., Boyle, S., Burtin, C., & Vogiatzis, I. (2019). Use of pedometers as a tool to promote daily physical activity levels in patients with COPD: A systematic review and meta-analysis. *European Respiratory Review, 28*(154), 190039. https://doi.org/10.1183/16000617.0039-2019

Baier, J. M., Funck, K. L., Vernstrøm, L., Laugesen, E., & Poulsen, P. L. (2021). Low physical activity is associated with impaired endothelial function in patients with type 2 diabetes and controls after 5 years of follow-up. *BMC Endocrine Disorders, 21*, 189. https://doi.org/10.1186/s12902-021-00857-9

Baillot, A., Black, M., Brunet, J., & Romain, A. J. (2020). Biopsychosocial correlates of physical activity and sedentary time in adults with severe obesity. *Clinical Obesity, 10*. https://doi.org/10.1111/cob.12355

Bakker, E. A., Hartman, Y. A. W., Hopman, M. T. E., et al. (2020). Validity and reliability of subjective methods to assess sedentary behaviour in adults: A systematic review and meta-analysis. *International Journal of Behavioral Nutrition and Physical Activity, 17*, 75. https://doi.org/10.1186/s12966-020-00972-1

Bell, J. A., et al. (2015). Healthy obesity and objective physical activity. *The American Journal of Clinical Nutrition, 102*, 268–275. https://doi.org/10.3945/ajcn.115.110924

Buono, E. W., Vrijens, B., Bosworth, H. B., Liu, L. Z., Zullig, L. L., & Granger, B. B. (2017). Coming full circle in the measurement of medication adherence: Opportunities and implications for health care. *Patient Preference and Adherence, 11*, 1009–1017. https://doi.org/10.2147/PPA.S127131

Burke, B. (1947). The dietary history as a tool in research. *Journal of the American Dietetic Association, 23*, 1041–1046. https://doi.org/10.1016/S0002-8223(21)43949-0

Buzzard, M. (1998). 24-hour dietary recall and food record methods. Cap 4. In W. Walter (Ed.), *Nutritional epidemiology* (2nd ed., pp. 50–73). Oxford University Press.

Camolas, J., Santos, O., Mascarenhas, M., Moreira, P., & Carmo, I. D. (2015). *INDIVÍDUO: intervenção nutricional direcionada aos estilos de vida em indivíduos com obesidade* (pp. 14–21). https://doi.org/10.21011/apn.2015.0303

Caspersen, C. J., Powell, K. E., & Christenson, G. M. (1985). Physical activity, exercise, and physical fitness: Definitions and distinctions for health-related research. *Public Health Reports, 100*(2), 126–121.

CDC – Centers for Disease Control and Prevention. (2021). *Type 2 Diabetes*. Centers for Disease Control and Prevention. https://www.cdc.gov/diabetes/basics/type2.html

Chapelot, D. (2011). The role of snacking in energy balance: A biobehavioral approach. *The Journal of Nutrition, 141*(1), 158–162. https://doi.org/10.3945/jn.109.114330

Chokshi, N. P., Adusumalli, S., Small, D. S., Morris, A., Feingold, J., Ha, Y. P., et al. (2018). Loss-framed financial incentives and personalized goal-setting to increase physical activity among ischemic heart disease patients using wearable devices: The ACTIVE REWARD randomized trial. *Journal of the American Heart Association, 7*. https://doi.org/10.1161/JAHA.118.009173

Cohen, J. L., Mann, D. M., Wisnivesky, J. P., Home, R., Leventhal, H., Musumeci-Szabó, T. J., & Halm, E. A. (2009). Assessing the validity of self-reported medication adherence among inner-city asthmatic adults: The medication adherence report scale for asthma. *Annals of Allergy, Asthma & Immunology : Official Publication of the American College of Allergy, Asthma, & Immunology, 103*(4), 325–331. https://doi.org/10.1016/s1081-1206(10)60532-7

Craig, C. L., Marshall, A. L., Sjöström, M., Bauman, A. E., Booth, M. L., Ainsworth, B. E., Pratt, M., Ekelund, U., Yngve, A., Sallis, J. F., & Oja, P. (2003). International physical activity questionnaire: 12-country reliability and validity. *Medicine and Science in Sports and Exercise, 35*(8), 1381–1395. https://doi.org/10.1249/01.MSS.0000078924.61453.FB

Critchley, J. A., & Capewell, S. (2003). Mortality risk reduction associated with smoking cessation in patients with coronary heart disease: A systematic review. *JAMA, 290*(1), 86–97. https://doi.org/10.1001/jama.290.1.86

Cryer, P. E. (2019). *Hypoglycemia in adults with diabetes mellitus*. In I. B. Hirsch & J. E. Mulder (Eds.), *UpToDate*. Retrieved 2019, June 5 from https://www.uptodate.com/contents/hypoglycemia-in-adults-with-diabetes-mellitus

DiCenso, A., Bayley, L., & Haynes, R. (2009). Accessing pre-appraised evidence: Fine-tuning the 5S model into a 6S model. *ACP Journal Club, 151*(3), 5–6. https://doi.org/10.1136/ebn.12.4.99-b

Dipietro, L., Caspersen, C. J., Ostfeld, A. M., & Nadel, E. R. (1993). A survey for assessing physical activity among older adults. *Medicine and Science in Sports and Exercise, 25*(5), 628–642.

Dowd, K. P., Szeklicki, R., Minetto, M. A., Murphy, M. H., Polito, A., Ghigo, E., et al. (2018). A systematic literature review of reviews on techniques for physical activity measurement in adults: A DEDIPAC study. *International Journal of Behavioral Nutrition and Physical Activity, 15*. https://doi.org/10.1186/s12966-017-0636-2

Dulloo, A. G., Jacquet, J., & Montani, J.-P. (2012). How dieting makes some fatter: From a perspective of human body composition autoregulation. *Proceedings of the Nutrition Society, 71*(03), 379–389. https://doi.org/10.1017/S0029665112000225

Esposito, K., Kastorini, C. M., Panagiotakos, D. B., & Giugliano, D. (2011). Mediterranean diet and weight loss: Meta-analysis of randomized controlled trials. *Metabolic Syndrome and Related Disorders, 9*(1), 1–12. https://doi.org/10.1089/met.2010.0031

Ezekowitz, J. A., O'Meara, E., McDonald, M. A., Abrams, H., Chan, M., Ducharme, A., et al. (2017). 2017 comprehensive update of the Canadian cardiovascular society guidelines for the management of heart failure. *Canadian Journal of Cardiology, 33*(11), 1342–1433. https://doi.org/10.1016/j.cjca.2017.08.022

Feig, E. H., Harnedy, L. E., Celano, C. M., & Huffman, J. C. (2021). Increase in daily steps during the early phase of a physical activity intervention for type 2 diabetes as a predictor of intervention outcome. *International Journal of Behavioral Medicine, 28*(6), 834–839. https://doi.org/10.1007/s12529-021-09966-0

Forbes, C. A., Deshpande, S., Sorio-Vilela, F., Kutikova, L., Duffy, S., Gouni-Berthold, I., & Hagström, E. (2018). A systematic literature review comparing methods for the measurement of patient persistence and adherence. *Current Medical Research and Opinion, 34*(9), 1613–1625. https://doi.org/10.1080/03007995.2018.1477747

Frühbeck, G., Busetto, L., Dicker, D., Yumuk, V., Goossens, G. H., Hebebrand, J., Halford, J., Farpour-Lambert, N. J., Blaak, E. E., Woodward, E., & Toplak, H. (2019). The ABCD of obesity: An EASO position statement on a diagnostic term with clinical and scientific implications. *Obesity Facts, 12*(2), 131–136. https://doi.org/10.1159/000497124

Fu, J., Liu, Y., Zhang, L., Zhou, L., Li, D., Quan, H., Zhu, L., Hu, F., Li, X., Meng, S., Yan, R., Zhao, S., Onwuka, J. U., Yang, B., Sun, D., & Zhao, Y. (2020). Nonpharmacologic interventions for reducing blood pressure in adults with prehypertension to established hypertension. *Journal of the American Heart Association, 9*(19), Article 19. https://doi.org/10.1161/JAHA.120.016804

GINA – Global Initiative for Asthma. (2021). *Global strategy for asthma management and prevention – Update 2021*. https://ginasthma.org/wp-content/uploads/2021/05/GINA-Main-Report-2021-V2-WMS.pdf

GOLD – Global Initiative for Chronic Obstructive Lung Disease. (2020). *Global strategy for prevention, diagnosis and management of COPD 2020 report*. https://goldcopd.org/gold-reports/

GOLD – Global Initiative for Chronic Obstructive Lung Disease. (2021). *Global strategy for prevention, diagnosis and management of COPD 2022 report*. https://goldcopd.org/2022-gold-reports-2/

Guerreiro, M. P., Strawbridge, J., Cavaco, A. M., Félix, I. B., Marques, M. M., & Cadogan, C. (2021). Development of a European competency framework for health and other professionals to support behaviour change in persons self-managing chronic disease. *BMC Medical Education, 21*(1), 1–14. https://doi.org/10.1186/s12909-021-02720-w

Hawks, L., Himmelstein, D. U., Woolhandler, S., Bor, D. H., Gaffney, A., & McCormick, D. (2020). Trends in unmet need for physician and preventive services in the United States, 1998–2017. *JAMA Internal Medicine, 180*, 439–444. https://doi.org/10.1001/jamainternmed.2019.6538

Hebert, J. R., Clemow, L., Pbert, L., Ockene, I. S., & Ockene, J. K. (1995). Social desirability bias in dietary self-report may compromise the validity of dietary intake measures. *International Journal of Epidemiology, 24*(2), 389–398. https://doi.org/10.1093/ije/24.2.389

Horne, R., & Weinman, J. (2002). Self-regulation and self-management in asthma: Exploring the role of illness perceptions and treatment beliefs in explaining non-adherence to preventer medication. *Psychology and Health., 17*(1), 17–32. https://doi.org/10.1080/08870440290001502

Hu, F., & Willett, W. (2018). Current and future landscape of nutritional epidemiologic research. *JAMA, 320*(20), 2073–2074. https://doi.org/10.1001/jama.2018.16166

Hussain, T., Awan, A. U., Abro, K. A., Ozair, M., & Manzoor, M. (2021). A mathematical and parametric study of epidemiological smoking model: A deterministic stability and optimality for solutions. *The European Physical Journal Plus, 136*(1), 1–23.

IDF – International Diabetes Federation. (2021). *IDF diabetes atlas 10th*. ISBN: 978-2-930229-98-0. Available at www.diabetesatlas.org.

Izawa, K. P., et al. (2014). Association between mental health and physical activity in patients with chronic heart failure. *Disability and Rehabilitation, 36*, 250–254. https://doi.org/10.310 9/09638288.2013.785604

Jaarsma, T., Arestedt, K. F., Mårtensson, J., Dracup, K., & Strömberg, A. (2009). The European heart failure self-care behaviour scale revised into a nine-item scale (EHFScB-9): A reliable and valid international instrument. *European Journal of Heart Failure, 11*(1), 99–105. https://doi.org/10.1093/eurjhf/hfn007

Jang, B. N., Kim, H. J., Kim, B. R., Woo, S., Lee, W. J., & Park, E. C. (2021). Effect of practicing health behaviors on unmet needs among patients with chronic diseases: A longitudinal study. *International Journal of Environmental Research and Public Health, 18*(15), 7977. https://doi.org/10.3390/ijerph18157977

Klompstra, L., Jaarsma, T., Strömberg, A., & van der Wal, M. H. L. (2019). Seasonal variation in physical activity in patients with heart failure. *Heart & Lung, 48*, 381–385. https://doi.org/10.1016/j.hrtlng.2019.06.004

Knuuti, J., Wijns, W., Saraste, A., Capodanno, D., Barbato, E., Funck-Brentano, C., et al. (2020). 2019 ESC guidelines for the diagnosis and management of chronic coronary syndromes: The task force for the diagnosis and management of chronic coronary syndromes of the European Society of Cardiology (ESC). *European Heart Journal, 41*(3), 407–477. https://doi.org/10.1093/eurheartj/ehz425

Krebs-Smith, S. M., Pannucci, T. E., Subar, A. F., Kirkpatrick, S. I., Lerman, J. L., Tooze, J. A., Wilson, M. M., & Reedy, J. (2018). Update of the healthy eating index: HEI-2015. *Journal of the Academy of Nutrition and Dietetics, 118*(9), 1591–1602. https://doi.org/10.1016/j.jand.2018.05.021

Lam, W. Y., & Fresco, P. (2015). Medication adherence measures: An overview. *BioMed Research International, 2015*(217047). https://doi.org/10.1155/2015/217047

Leischik, R., Schwarz, K., Bank, P., Brzek, A., Dworrak, B., Strauss, M., et al. (2021). Exercise improves cognitive function—A randomized trial on the effects of physical activity on cognition in type 2 diabetes patients. *Journal of Personalized Medicine, 11*, 530. https://doi.org/10.3390/jpm11060530

Liguori, G. (Ed.). (2021). *ACSM's guidelines for exercise testing and prescription* (11th ed.). EUA: Wolters Kluwer Health.

Lopes, D., Ribeiro, I. S., Santos, D. C., Lima, F., Santos, A. A., Souza, D., Lopes, D. N., Prado, A. O., Pereira, Í. S., Santos, D. P., Santos, G. S., & Silva, R. (2021). Regular physical activity reduces the proinflammatory response in older women with diabetes and hypertension in the postmenopausal phase. *Experimental Gerontology, 152*, 111449. https://doi.org/10.1016/j.exger.2021.111449

Maddison, R., Pfaeffli, L., Whittaker, R., Stewart, R., Kerr, A., Jiang, Y., Kira, G., Leung, W., Dalleck, L., Carter, K., & Rawstorn, J. (2015). A mobile phone intervention increases physical activity in people with cardiovascular disease: Results from the HEART randomized controlled trial. *European Journal of Preventive Cardiology, 22*(6), 701–709. https://doi.org/10.1177/2047487314535076

Marcos-Pardo, P. J., González-Gálvez, N., López-Vivancos, A., Espeso-García, A., Martínez-Aranda, L. M., Gea-García, G. M., Orquín-Castrillón, F. J., Carbonell-Baeza, A., Jiménez-García, J. D., Velázquez-Díaz, D., Cadenas-Sanchez, C., Isidori, E., Fossati, C., Pigozzi, F., Rum, L., Norton, C., Tierney, A., Äbelkalns, I., Klempere-Sipjagina, A., Porozovs, J., et al. (2020). Sarcopenia, diet, physical activity and obesity in European middle-aged and older adults: The LifeAge study. *Nutrients, 13*(1), 8. https://doi.org/10.3390/nu13010008

Martinez Aguirre-Betolaza, A., Mujika, I., Loprinzi, P., Corres, P., Gorostegi-Anduaga, I., & Maldonado-Martín, S. (2020). Physical activity, sedentary behavior, and sleep quality in adults with primary hypertension and obesity before and after an aerobic exercise program: EXERDIET-HTA study. *Life (Basel, Switzerland), 10*(8), 153. https://doi.org/10.3390/life10080153

McDonagh, T. A., Metra, M., Adamo, M., Gardner, R. S., Baumbach, A., Böhm, M., et al. (2021). 2021 ESC guidelines for the diagnosis and treatment of acute and chronic heart failure: Developed by the task force for the diagnosis and treatment of acute and chronic heart failure of the European Society of Cardiology (ESC) with the special contribution of the Heart Failure Association (HFA) of the ESC. *European Heart Journal, 42*(36), 3599–3726. https://doi.org/10.1002/ejhf.2333

Mihaltan, F., Adir, Y., Antczak, A., Porpodis, K., Radulovic, V., Pires, N., de Vries, G. J., Horner, A., De Bontridder, S., Chen, Y., Shavit, A., Alecu, S., & Adamek, L. (2019). Importance of the relationship between symptoms and self-reported physical activity level in stable COPD based on the results from the SPACE study. *Respiratory Research, 20*(1), 89. https://doi.org/10.1186/s12931-019-1053-7

Miller, T. M., Abdel-Maksoud, M. F., Crane, L. A., Marcus, A. C., & Byers, T. E. (2008). Effects of social approval bias on self-reported fruit and vegetable consumption: A randomized controlled trial. *Nutrition Journal, 7*, 18. https://doi.org/10.1186/1475-2891-7-18

Morisky, D. E., Ang, A., Krousel-Wood, M., & Ward, H. J. (2008). Predictive validity of a medication adherence measure in an outpatient setting. *Journal of Clinical Hypertension (Greenwich, Conn.), 10*(5), 348–354. https://doi.org/10.1111/j.1751-7176.2008.07572.x

Morita, A. A., Silva, L., Bisca, G. W., Oliveira, J. M., Hernandes, N. A., Pitta, F., & Furlanetto, K. C. (2018). Heart rate recovery, physical activity level, and functional status in subjects with COPD. *Respiratory Care, 63*(8), 1002–1008. https://doi.org/10.4187/respcare.05918

Panizza, C. E., Shvetsov, Y. B., Harmon, B. E., Wilkens, L. R., Le Marchand, L., Haiman, C., Reedy, J., & Boushey, C. J. (2018). Testing the predictive validity of the healthy eating Index-2015 in the multiethnic cohort: Is the score associated with a reduced risk of all-cause and cause-specific mortality. *Nutrients, 10*(4), 452. https://doi.org/10.3390/nu10040452

Pedersen, B. K., & Saltin, B. (2015). Exercise as medicine – Evidence for prescribing exercise as therapy in 26 different chronic diseases. *Scandinavian Journal of Medicine & Science in Sports, 25*(Suppl 3), 1–72. https://doi.org/10.1111/sms.12581

Piercy, K. L., Troiano, R. P., Ballard, R. M., Carlson, S. A., Fulton, J. E., Galuska, D. A., George, S. M., & Olson, R. D. (2018). The physical activity guidelines for Americans. *JAMA, 320*(19), 2020–2028. https://doi.org/10.1001/jama.2018.14854

Pinto, T. F., Fagundes Xavier, R., Lunardi, A. C., Marques da Silva, C., Moriya, H. T., Lima Vitorasso, R., Torsani, V., Amato, M., Stelmach, R., Salge, J. M., Carvalho-Pinto, R. M., & Carvalho, C. (2020). Effects of elastic tape on thoracoabdominal mechanics, dyspnea, exercise capacity, and physical activity level in nonobese male subjects with COPD. *Journal of Applied Physiology (Bethesda, Md. : 1985), 129*(3), 492–499. https://doi.org/10.1152/japplphysiol.00690.2019

Piopoli, M. F., Corrà, U., Adamopoulos, S., Benzer, W., Bjarnason-Wehrens, B., Cupples, M., Dendale, P., Doherty, P., Gaita, D., Höfer, S., McGee, H., Mendes, M., Niebauer, J., Pogosova, N., Garcia-Porrero, E., Rauch, B., Schmid, J. P., & Giannuzzi, P. (2014). Secondary prevention in the clinical management of patients with cardiovascular diseases. Core components, standards and outcome measures for referral and delivery: A policy statement from the cardiac rehabilitation section of the European Association for Cardiovascular Prevention & rehabilitation. *Endorsed by the Committee for Practice Guidelines of the European Society of Cardiology. European Journal of Preventive Cardiology, 21*(6), 664–681. https://doi.org/10.1177/2047487312449597

Pound, C. M., Zhang, J. Z., Kodua, A. T., & Sampson, M. (2021). Smoking cessation in individuals who use vaping as compared with traditional nicotine replacement therapies: A systematic review and meta-analysis. *BMJ Open, 11*(2), e044222. https://doi.org/10.1136/bmjopen-2020-044222

Ramirez-Marrero, F. A., Miles, J., Joyner, M. J., & Curry, T. B. (2014). Self-reported and objective physical activity in postgastric bypass surgery, obese and lean adults: Association with body composition and cardiorespiratory fitness. *Journal of Physical Activity & Health, 11*, 145–151. https://doi.org/10.1123/jpah.2012-0048

Ratamess, N. A. (2021). *ACSM's foundations of strength training and conditioning* (2nd ed.). Wolters Kluwer Health.

Reedy, J., Lerman, J. L., Krebs-Smith, S. M., Kirkpatrick, S. I., Pannucci, T. R. E., Wilson, M. M., Subar, A. F., Kahle, L. L., & Tooze, J. A. (2018). Evaluation of the healthy eating Index-2015. *Journal of the Academy of Nutrition and Dietetics, 118*(9), 1622–1633.

Reid, R. D., Morrin, L. I., Beaton, L. J., Papadakis, S., Kocourek, J., McDonnell, L., Slovinec D'Angelo, M. E., Tulloch, H., Suskin, N., Unsworth, K., Blanchard, C., & Pipe, A. L. (2012). Randomized trial of an internet-based computer-tailored expert system for physical activity in patients with heart disease. *European Journal of Preventive Cardiology, 19*(6), 1357–1364. https://doi.org/10.1177/1741826711422988

Riegel, G. R., Martins, G. B., Schmidt, A. G., Rodrigues, M. P., Nunes, G. S., Correa, V., Jr., Fuchs, S. C., Fuchs, F. D., Ribeiro, P. A., & Moreira, L. B. (2019). Self-reported adherence to physical activity recommendations compared to the IPAQ interview in patients with hypertension. *Patient Preference and Adherence, 13*, 209–214. https://doi.org/10.2147/PPA.S185519

Rodrigues, B., Encantado, J., Carraça, E., Sousa-Sá, E., Lopes, L., Cliff, D., Mendes, R., Silva, M. N., Godinho, C., & Santos, R. (2022). Questionnaires measuring movement behaviours in adults and older adults: Content description and measurement properties. A systematic review. *PLOS One, 17*(3), e0265100. https://doi.org/10.1371/journal.pone.0265100

Rossen, J., Larsson, K., Hagströmer, M., Yngve, A., Brismar, K., Ainsworth, B., Åberg, L., & Johansson, U. B. (2021). Effects of a three-armed randomised controlled trial using self-monitoring of daily steps with and without counselling in prediabetes and type 2 diabetes-the Sophia Step Study. *The International Journal of Behavioral Nutrition and Physical Activity, 18*(1), 121. https://doi.org/10.1186/s12966-021-01193-w

Sallis, J. F., Haskell, W. L., Wood, P. D., Fortmann, S. P., Rogers, T., Blair, S. N., & Paffenbarger, R. S. (1985). Physical activity assessment methodology in the Five-City project. *American Journal of Epidemiology, 121*(1), 91–106.

Schlenk, E. A., Fitzgerald, G. K., Rogers, J. C., Kwoh, C. K., & Sereika, S. M. (2021). Promoting physical activity in older adults with knee osteoarthritis and hypertension: A randomized controlled trial. *Journal of Aging and Physical Activity, 29*, 207–218. https://doi.org/10.1123/japa.2019-0498

Schmidt, C., Santos, M., Bohn, L., Delgado, B. M., Moreira-Gonçalves, D., Leite-Moreira, A., & Oliveira, J. (2020). Comparison of questionnaire and accelerometer-based assessments of physical activity in patients with heart failure with preserved ejection fraction: Clinical and prognostic implications. *Scandinavian Cardiovascular Journal: SCJ, 54*(2), 77–83. https://doi.org/10.1080/14017431.2019.1707863

Schröder, H., Fitó, M., Estruch, R., Martínez-González, M. A., Corella, D., Salas-Salvadó, J., Lamuela-Raventós, R., Ros, E., Salaverría, I., Fiol, M., Lapetra, J., Vinyoles, E., Gómez-Gracia, E., Lahoz, C., Serra-Majem, L., Pintó, X., Ruiz-Gutierrez, V., & Covas, M. I. (2011). A short screener is valid for assessing Mediterranean diet adherence among older Spanish men and women. *The Journal of Nutrition, 141*(6), 1140–1145. https://doi.org/10.3945/jn.110.135566

Schutz, D. D., Busetto, L., Dicker, D., Farpour-Lambert, N., Pryke, R., Toplak, H., Widmer, D., Yumuk, V., & Schutz, Y. (2019). European practical and patient-centred guidelines for adult obesity management in primary care. *Obesity Facts, 12*(1), 40–66. https://doi.org/10.1159/000496183

Silbert, R., Salcido-Montenegro, A., Rodriguez-Gutierrez, R., Katabi, A., & McCoy, R. G. (2018). Hypoglycemia among patients with type 2 diabetes: Epidemiology, risk factors, and prevention strategies. *Current Diabetes Reports, 18*(8), 53. https://doi.org/10.1007/s11892-018-1018-0

Sousa Junior, A. E., Macêdo, G., Schwade, D., Sócrates, J., Alves, J. W., Farias-Junior, L. F., Freire, Y. A., Lemos, T., Browne, R., & Costa, E. C. (2020). Physical activity counseling for adults with hypertension: A randomized controlled pilot trial. *International Journal of Environmental Research and Public Health, 17*(17), 6076. https://doi.org/10.3390/ijerph17176076

Svarstad, B. L., Chewning, B. A., Sleath, B. L., & Claesson, C. (1999). The brief medication questionnaire: A tool for screening patient adherence and barriers to adherence. *Patient Education and Counseling, 37*(2), 113–124. https://doi.org/10.1016/s0738-3991(98)00107-4

Tasnim, S., Tang, C., Musini, V. M., & Wright, J. M. (2020). Effect of alcohol on blood pressure. *Cochrane Database of Systematic Reviews, 7*. https://doi.org/10.1002/14651858. CD012787.pub2

Tesfaye, N., & Seaquist, E. R. (2010). Neuroendocrine responses to hypoglycemia. *Annals of the New York Academy of Sciences, 1212*, 12–28. https://doi.org/10.1111/j.1749-6632.2010.05820.x

Toobert, D. J., Hampson, S. E., & Glasgow, R. E. (2000). The summary of diabetes self-care activities measure: Results from 7 studies and a revised scale. *Diabetes Care, 23*(7), 943–950. https://doi.org/10.2337/diacare.23.7.943

Tudor-Locke, C., Craig, C. L., Aoyagi, Y., Bell, R. C., Croteau, K. A., De Bourdeaudhuij, I., Ewald, B., Gardner, A. W., Hatano, Y., Lutes, L. D., Matsudo, S. M., Ramirez-Marrero, F. A., Rogers, L. Q., Rowe, D. A., Schmidt, M. D., Tully, M. A., & Blair, S. N. (2011a). How many steps/day are enough? For older adults and special populations. *The International Journal of Behavioral Nutrition and Physical Activity, 8*, 80. https://doi.org/10.1186/1479-5868-8-80

Tudor-Locke, C., Leonardi, C., Johnson, W. D., Katzmarzyk, P. T., & Church, T. S. (2011b). Accelerometer steps/day translation of moderate-to-vigorous activity. *Preventive Medicine, 53*, 31–33. https://doi.org/10.1016/j.ypmed.2011.01.014

West, R., Hajek, P., Stead, L., & Stapleton, J. (2005). Outcome criteria in smoking cessation trials: Proposal for a common standard. *Addiction (Abingdon, England), 100*(3), 299–303. https://doi.org/10.1111/j.1360-0443.2004.00995.x

Whelton, P. K., Carey, R. M., Aronow, W. S., Casey, D. E., Collins, K. J., Himmelfarb, C. D., DePalma, S. M., Gidding, S., Jamerson, K. A., Jones, D. W., MacLaughlin, E. J., Muntner, P., Ovbiagele, B., Smith, S. C., Spencer, C. C., Stafford, R. S., Taler, S. J., Thomas, R. J., Williams, K. A., et al. (2018). 2017 ACC/AHA/AAPA/ABC/ACPM/AGS/APhA/ASH/ASPC/NMA/PCNA guideline for the prevention, detection, evaluation, and Management of High Blood Pressure in adults: Executive summary: A report of the American College of Cardiology/American Heart Association Task Force on Clinical Practice Guidelines. *Hypertension, 71*(6), 1269–1324. https://doi.org/10.1161/HYP.0000000000000066

WHO – World Health Organization. (2003). Diet, nutrition and the prevention of chronic diseases. World Health Organization technical report series, 916.

WHO (2005). *Global Physical Activity Questionnaire (GPAQ). WHO STEPwise approach to NCD risk factor surveillance*. World Health Organization: Prevention of Noncommunicable Diseases Department. Switzerland.

WHO – World Health Organization. (2015). *Guideline: Sugars intake for adults and children*. World Health Organization.

WHO (2020). *WHO guidelines on physical activity and sedentary behaviour*. World Health Organization. Switzerland. ISBN 978-92-4-001512-8 (electronic version). https://www.who.int/publications/i/item/9789240015128

Willet, W. (1990). In W. Willet (Ed.), *Nutritional epidemiology*. Oxford University Press.

Williams, B., Mancia, G., Spiering, W., Agabiti Rosei, E., Azizi, M., Burnier, M., et al. (2018). 2018 ESC/ESH guidelines for the management of arterial hypertension: The task force for the management of arterial hypertension of the European Society of Cardiology (ESC) and the European Society of Hypertension (ESH). *European Heart Journal, 39*(33), 3021–3104. https://doi.org/10.1080/08037051.2018.1527177

Witschi, J. C. (1990). Short-term dietary recall and recording methods. In W. Willet (Ed.), *Nutritional epidemiology* (pp. 52–68). Oxford University Press.

Yates, B. C., Pozehl, B., Kupzyk, K., Epstein, C. M., & Deka, P. (2017). Are heart failure and coronary artery bypass surgery patients meeting physical activity guidelines? *Rehabilitation Nursing, 42*(3), 119–124. https://doi.org/10.1002/rnj.257

Ye, F., Yale, S., Zheng, Y., Hu, H., Zhou, L., Fanning, J., Yeboah, J., Brubaker, P., & Bertoni, A. G. (2021). Trends in physical activity among US adults with heart failure, 2007–2016. *Journal of Cardiopulmonary Rehabilitation and Prevention, 41*(5), 351–356. https://doi. org/10.1097/HCR.0000000000000578

Yumuk, V., Tsigos, C., Fried, M., Schindler, K., Busetto, L., Micic, D., & Toplak, H. (2015). European guidelines for obesity management in adults. *Obesity Facts, 8*(6), 402–424. https:// doi.org/10.1159/000442721

Open Access   This chapter is licensed under the terms of the Creative Commons Attribution 4.0 International License (http://creativecommons.org/licenses/by/4.0/), which permits use, sharing, adaptation, distribution and reproduction in any medium or format, as long as you give appropriate credit to the original author(s) and the source, provide a link to the Creative Commons license and indicate if changes were made.

The images or other third party material in this chapter are included in the chapter's Creative Commons license, unless indicated otherwise in a credit line to the material. If material is not included in the chapter's Creative Commons license and your intended use is not permitted by statutory regulation or exceeds the permitted use, you will need to obtain permission directly from the copyright holder.

Chapter 4
Implementing Behaviour Change Strategies

Isa Brito Félix and Mara Pereira Guerreiro

Learning Outcomes

This chapter contributes to achieving the following learning outcomes:

BC3.1 Identify standardised sources of behaviour change techniques (BCTs).

BC3.2 Identify core BCTs for the self-management of chronic disease.

BC3.3A Provide examples of determinants in selected target behaviours.

BC3.3 Explain how behaviour determinants (opportunities and barriers) influence the selection of BCTs.

BC3.4 Apply core and supplementary BCTs in different target behaviours.

BC4.3 Discuss health behaviour determinants in light of clinical hallmarks, progression and complications of chronic diseases.

BC8.2 Demonstrate how to assess behaviour determinants through structured questionnaires, interview and other approaches.

BC11.1 Demonstrate critical understanding of BCTs appropriate for brief or long-term behaviour interventions.

I. B. Félix (✉)
Nursing Research, Innovation and Development Centre of Lisbon (CIDNUR),
Nursing School of Lisbon, Lisbon, Portugal
e-mail: isabsfelix@gmail.com

M. P. Guerreiro
Nursing Research, Innovation and Development Centre of Lisbon (CIDNUR),
Nursing School of Lisbon, Lisbon, Portugal

Egas Moniz Interdisciplinary Research Center (CiiEM), Egas Moniz School of
Health & Science, Monte de Caparica, Portugal

© The Author(s) 2023
M. P. Guerreiro et al. (eds.), *A Practical Guide on Behaviour Change Support
for Self-Managing Chronic Disease*, https://doi.org/10.1007/978-3-031-20010-6_4

4.1 Opportunities and Barriers to Implementing Change in Target Behaviours

4.1.1 Behaviour Determinants

Behaviour is influenced by determinants, as explained in Chap. 2. A key consideration is changeability, i.e. the extent to which determinants can be changed and the impact of those changes in influencing the target behaviour (Hankonen & Hardeman, 2020). Changeable factors that have a strong relationship to the behaviour are potential targets for interventions (Michie et al., 2011), impacting on intervention success (Williams et al., 2019).

Unmodifiable determinants are those that are unchangeable by a behavioural intervention, such as age. They may, however, influence the choice of an appropriate intervention. For instance, unemployment may have a negative influence on physical activity; although this barrier is not amenable to change by a behavioural intervention, it may be useful to tailor it (e.g. recommending strategies that do not involve spending money).

Chapter 3 presented key behaviours for self-management of high-priority chronic diseases, which may be influenced by a plethora of determinants. Examples are provided below for each target behaviour using the COM-B model, presented in Chap. 2. These examples do not intend to be exhaustive; they were collated based on case studies developed in the Train4Health project and the literature. An important consideration is that each person presents a unique combination of behaviour determinants based on morbidities, functional status, activities of daily living, preferences, resources and context. For example, forgetfulness may be a barrier to medication-taking in one person, while for others, not taking the medication may be related to concerns with side effects.

Similar barriers and facilitators may be observed across different target behaviours. For instance, facilitators for healthy diet may include social support or perceived self-efficacy, also identified as facilitators for physical activity.

4.1.1.1 Diet Including Alcohol Intake

Diet and alcohol intake are influenced by the interplay of behavioural, emotional and social factors, in addition to neuroendocrine and genetic influences. Certain religions limit the alcohol use, which can be seen as a facilitator within social opportunity (Kelly et al., 2018), according to the COM-B model. Influence of drinking alcohol habits of spouse/partner/family members/peers (Kelly et al., 2018) is also linked to social opportunity (Kelly et al., 2018). Examples of diet barriers and facilitators are shown in Table 4.1.

4.1.1.2 Physical Activity

Determinants for physical activity behaviour are presented in Table 4.2; examples include work schedule, social support, economic circumstances and energy.

4.1.1.3 Smoking Cessation

Barriers in smoking cessation include systems, organisations and the relationship between systems and individuals, for example, lack of access to smoking cessation programmes. Individual factors also influence quitting smoking such as physical addiction to nicotine. A range of most reported barriers to smoking cessation can be found in literature, such as enjoyment, craving, stress management and withdrawal symptoms. Common smoking cessation determinants are organised in Table 4.3.

Table 4.1 Examples of diet determinants

COM-B (West & Michie, 2020)		Determinant	
		Barriers	Facilitators
Capability *An attribute of a person that together with opportunity makes a behaviour possible or facilitates it*	**Physical capability** *Capability that involves a person's physique and musculoskeletal functioning (e.g. balance and dexterity)*	Tiredness to cook	Good cooking skills
		Physical disability to cook	
	Psychological capability *Capability that involves a person's mental functioning (e.g. understanding and memory)*	Lack of knowledge (e.g. a person with diabetes that does not know why and how to improve his/ her diet)	Planning either to purchase food at work or prepare in advance food to bring to work
		Perception of time constraints	Being able to do a grocery list
		Lack of monitoring of food consumption	Being able to understand quantities
Motivation *An aggregate of mental processes that energise and direct behaviour*	**Reflective motivation** *Motivation that involves conscious thought processes (e.g. plans and evaluations)*	Lack of motivation	Perceived confidence in ability to cook
		Discouragement due to lack of results (Cheng et al., 2019)	
		Pleasure with eating foods containing added sugars and fats	
	Automatic motivation *Motivation that involves habitual, instinctive, drive- related and affective processes (e.g. desires and habits)*	Depression which may lead to self-neglect	Habit of eating vegetables, unsweetened cereals and fruits
		Anxiety or stress (e.g. leading to eat comfort food and snacks)	

(continued)

Table 4.1 (continued)

COM-B (West & Michie, 2020)		Determinant	
		Barriers	Facilitators
Opportunity *An attribute of an environmental system that together with capability makes a behaviour possible or facilitates it*	**Physical opportunity** *Opportunity that involves inanimate parts of the environmental system and time (e.g. financial and material resources)*	Time constraints	Ability to grow and produce food (Seguin et al., 2014)
		Accessibility and availability of unhealthy options (e.g. easy access to goodies)	Accessibility to farmers' markets and farm shares (Seguin et al., 2014)
		Price of healthy foods (Cradock et al., 2021; Pinho et al., 2018)	Having only healthy choices to eat at home
		Lack of healthy options (Pinho et al., 2018)	Availability of health eating in the local restaurant or café (Cradock et al., 2021)
	Social opportunity *Opportunity that involves other people and organisations (e.g. culture and social norms)*	Social context (e.g. inviting people to one's home and serving food and drink, enjoying drinking alcohol with friends)	Positive influences of family and friends on healthy eating behaviour (Cheng et al., 2019; Cradock et al., 2021)
		Taste preferences of family and friends (e.g. people around eating confectionery)	Religion

Table 4.2 Examples of physical activity determinants

COM-B (West & Michie, 2020)		Determinant	
		Barriers	**Facilitator**
Capability *An attribute of a person that together with opportunity makes a behaviour possible or facilitates it*	**Physical capability** *Capability that involves a person's physique and musculoskeletal functioning (e.g. balance and dexterity)*	Fatigue (Cortis et al., 2017)	Physical health status (e.g. a person who has lost weight may feel more energetic to engage in physical activity)
		Health status change (e.g. worsening dyspnoea in a person with COPD)	
	Psychological capability *Capability that involves a person's mental functioning (e.g. understanding and memory)*	Lack of knowledge about the importance of physical activity	Knowledge about the health consequences of an inactive lifestyle
			Physical activity planning (e.g. having a routine to do physical activity)

(continued)

Table 4.2 (continued)

| COM-B (West & Michie, 2020) | | Determinant | |
		Barriers	**Facilitator**
Motivation *An aggregate of mental processes that energise and direct behaviour*	**Reflective motivation** *Motivation that involves conscious thought processes (e.g. plans and evaluations)*	Lack of prioritisation assigned to physical activity	Motivation to adopt and maintain healthy physical activity behaviour
		Preferences for sedentary activities at home (e.g. reading or watching television)	High levels of self-efficacy (e.g. perceived confidence in the ability to run 3 km)
		Lack of care with body image	Enjoyment (Cortis et al., 2017)
		Fear of injury (Cortis et al., 2017)	
Opportunity *An attribute of an environmental system that together with capability makes a behaviour possible or facilitates it*	**Physical opportunity** *Opportunity that involves inanimate parts of the environmental system and time (e.g. financial and material resources)*	Heavy work schedule	Good weather (e.g. when physical activities are performed outdoors)
		Lack of time (e.g. busywork routines or family obligations) (Cradock et al., 2021)	
		Economic circumstances (e.g. being unable to pay for the gym or group classes) (Cradock et al., 2021)	
		Neighbourhood (e.g. lack of parks or sidewalks or open spaces; long distance from parks)	
	Social opportunity *Opportunity that involves other people and organisations (e.g. culture and social norms)*	Lack of social support (e.g. no encouragement from family and friends to perform physical activity) (Cortis et al., 2017)	Social support (e.g. a friend that helps the person to keep focused and motivated)

Table 4.3 Examples of smoking cessation determinants

| COM-B (West & Michie, 2020) | | Determinant | |
		Barriers	**Facilitator**
Capability *An attribute of a person that together with opportunity makes a behaviour possible or facilitates it*	**Physical capability** *Capability that involves a person's physique and musculoskeletal functioning (e.g. balance and dexterity)*	Nicotine addiction (Chean et al., 2019)	
		Withdrawal symptoms on quitting (Chean et al., 2019)	
	Psychological capability *Capability that involves a person's mental functioning (e.g. understanding and memory)*	Lack of knowledge about smoking cessation consultations	Planning skills for seeking smoking cessation help
Motivation *An aggregate of mental processes that energise and direct behaviour*	**Reflective motivation** *Motivation that involves conscious thought processes (e.g. plans and evaluations)*	Negative impression about the effectiveness of assisted smoking cessation (Chean et al., 2019)	High level of self-efficacy
		Pleasure from smoking	Willingness to quit smoking
	Automatic motivation *Motivation that involves habitual, instinctive, drive-related and affective processes (e.g. desires and habits)*	Stress/anxiety (Ferra et al., 2019)	
		Impaired capacity for self-control (Chean et al., 2019)	
		Impulse (i.e. the decision to resume smoking is rather impulsive)	
Opportunity *An attribute of an environmental system that together with capability makes a behaviour possible or facilitates it*	**Physical opportunity** *Opportunity that involves inanimate parts of the environmental system and time (e.g. financial and material resources)*	Unaffordable smoking cessation medication (Ferra et al., 2019)	Affordable smoking cessation medication (Ferra et al., 2019)
		Easy access to cigarettes (Ferra et al., 2019)	Restricted access to tobacco (Ferra et al., 2019)
		Limited access to smoking cessation programmes (Ferra et al., 2019)	Easy access to smoking cessation programmes (Ferra et al., 2019)
			Smoke-free homes and places
	Social opportunity *Opportunity that involves other people and organisations (e.g. culture and social norms)*	Example from others (e.g. friends who smoke in social activities or workplaces) (Chean et al., 2019)	Social support from family and friend to quit smoking
		Cigarette offers from friends and relatives (Chean et al., 2019)	

4.1.1.4   Medication Adherence

Medication adherence is affected by multiple determinants such as psychosocial, economic and health system factors (Kardas et al., 2013; Kvarnström et al., 2021; Mishra et al., 2021). For instance, a strong network providing social support increases medication adherence, while forgetfulness may contribute to non-adherence (Kardas et al., 2013; Kvarnström et al., 2021). Table 4.4 provides examples of barriers and facilitators of medication adherence.

Table 4.4 Examples of medication adherence determinants

COM-B (West & Michie, 2020)		Determinant	
		Barriers	**Facilitator**
Capability *An attribute of a person that together with opportunity makes a behaviour possible or facilitates it*	**Physical capability** *Capability that involves a person's physique and musculoskeletal functioning (e.g. balance and dexterity)*	Lack of dexterity to take the medication	Planning medication taken
		Unplanned travel or routine changes (Kvarnström et al., 2021)	Knowledge about prescribed medication
	Psychological capability *Capability that involves a person's mental functioning (e.g. understanding and memory)*	Forgetfulness	Integrating meditation into daily life (Kvarnström et al., 2021)
		Incapability of planning medication-taking	
Motivation *An aggregate of mental processes that energise and direct behaviour*	**Reflective motivation** *Motivation that involves conscious thought processes (e.g. plans and evaluations)*	Beliefs about lack of necessity (e.g. these medicines don't protect me from becoming worse; my health, at present, does not depend on these medicines, adapted from Horne et al., 1999) (Félix & Henriques, 2021)	Perception of disease severity (Kardas et al., 2013; Kvarnström et al., 2021)
		Concerns about medication (e.g. these medicines give me unpleasant side effects; these medicines disrupt my life) (Félix & Henriques, 2021; Horne et al., 1999)	Fear of recurrence of event, (e.g. physical pain and the fear of recurrence (Mishra et al., 2021)
	Automatic motivation *Motivation that involves habitual, instinctive, drive-related and affective processes (e.g. desires and habits)*	Depression (Félix & Henriques, 2021; Jackson et al., 2014; Kardas et al., 2013)	Habit

(continued)

Table 4.4 (continued)

COM-B (West & Michie, 2020)		Determinant	
		Barriers	**Facilitator**
Opportunity *An attribute of an environmental system that together with capability makes a behaviour possible or facilitates it*	**Physical opportunity** *Opportunity that involves inanimate parts of the environmental system and time (e.g. financial and material resources)*	Lack of medication availability (Kvarnström et al., 2021)	Good access to a healthcare facility (Kardas et al., 2013)
		Cost of medication	
		Lack of health insurance (Mishra et al., 2021)	
		Lack of clarity in prescription (Mishra et al., 2021)	
	Social opportunity *Opportunity that involves other people and organisations (e.g. culture and social norms)*	Cultural preference for alternative medicine (Kvarnström et al., 2021)	Support from healthcare professionals
		Lack of social support (Kardas et al., 2013)	Emotional or practical support by family members or careers (Kardas et al., 2013; Kvarnström et al., 2021; Mishra et al., 2021)

Clinical hallmarks, progression and complications of chronic diseases should be considered, as they may influence self-management behaviours. For example, the progression of COPD and the existence of dyspnoea on exertion may negatively influence physical activity. Furthermore, it may also be directly related to diet behaviour (e.g. if the person does not have the capability to go to the supermarket frequently due to fatigue, eating healthy food may be compromised).

Another example is a person who had a leg amputation due to type 2 diabetes complications and does not have a prosthesis; this may represent a barrier to specific exercises or physical activity. In the case of retinopathy caused by type 2 diabetes, recognising medicines may become difficult, which may influence how people take them.

In summary, clinical characteristics, progression and complications of chronic diseases, as well as other determinants exemplified previously should be assessed when planning a behaviour change intervention.

4.1.2   Assessing Behaviour Determinants Using Appropriate Measures

The previous section illustrates different determinants that influence target behaviours in the self-management of chronic disease. As already explained, the examples presented do not intend to be exhaustive. While these can be helpful to bear in mind when assessing behaviour determinants, it is equally important not to forget that each person is unique and can present specific barriers and facilitators.

Identifying the key determinants that influence a target behaviour often requires a range of methods and sources. This section summarises approaches to assess determinants in order to guide the intervention plan.

The interview is the most frequent approach in practice to assess behaviour determinants. As explained in Chap. 5 "Communication and Person-Centred Behaviour Change", the interview ideally starts with open questions, to expand the dialogue and unravel the person's perspective on barriers and facilitators, and can then move to closed-ended questions, to fine-tune the understanding and clarify details. Possible questions are presented below (Boxes 4.1 and 4.2).

Box 4.1 Example of Open Questions
- Why have you been having difficulty increasing your fruit consumption?
- What helps you to take your medication/to increase your physical activity?
- What do you think is needed to eat fewer carbs and more veggies?
- What thoughts have you had about increasing walking?
- What is your biggest barrier to stop smoking?

Box 4.2 Example of Closed-Ended Questions
- Do you feel confident about managing your medication?
- Does the cost of healthy foods influence your behaviour?
- Do stress levels make you crave for a cigarette?
- Does pain in your knees bother you when walking?

The interview can be supplemented with tools to assess determinants; some tools are behaviour-specific (Table 4.5), while others are disease-specific and ascertain determinants in more than one target behaviour. The Diabetes Self-Efficacy scale is one example of a Likert-type scale with eight items that assess self-efficacy in different target behaviours, such as diet, physical activity and medication-taking (Ritter et al., 2016).

Keyworth et al., (2020) developed a novel six-item questionnaire for self-evaluating people's perceptions of capabilities, opportunities and motivations based on the COM-B model. This questionnaire is sufficiently generic for any behaviour or population. Respondents rate the level of agreement with the statement (e.g. I have the physical opportunity to *change my behaviour to improve my health*) on a scale from 0 to 10. Alternative text is presented in italics and can be replaced by a target behaviour such as physical activity or diet. Then, specific barriers or facilitators of behaviour can be explored for each COM-B component.

Table 4.5 Examples of behaviour-specific tools to assess determinants

Target behaviour	Determinant	Tool
Medication adherence	Beliefs about medicines	*Beliefs About Medicines Questionnaire – Specific (BMQ-Specific)* (Horne et al., 1999) The BMQ-specific is comprised of two five-item subscales to assess the person's beliefs about the necessity of prescribed medication and their concerns about potential adverse consequences of taking it: the necessity and the concerns scale. Each item is scored on a five-point Likert scale that varies from strongly disagree (1) to strongly agree (5). Total scores for the necessity and concerns scales are obtained by summing the scores of individual items (min. 5 and max. 25). The higher the score is, the greater the person's belief in the construct
	Multiple determinants	*Identification of Medication Adherence Barriers Questionnaire (IMAB-Q 30 or IMAB-Q 10)* (Brown et al., 2017) The IMAB-Q 30 and IMAB-Q 10 address potential barriers to medication adherence. IMAB-Q 10 is a 10-item questionnaire, while the IMAB-Q 30 comprises 30 items. Each item corresponds to a barrier that may be experienced on medication-taking by a person. Both questionnaires have a five-point Likert scale ranging from strongly agree to strongly disagree
	Self-efficacy	*The Self-Efficacy for Appropriate Medication Use Scale (SEAMS)* (Risser et al., 2007) The SEAMS is a 13-item scale to assess medication self-efficacy in chronic disease management. This three-point response scale – (1) not confident, (2) somewhat confident and (3) very confident – measures the level of confidence of taking medication as prescribed in different scenarios. An example of an item is "how confident are you that you can take your medicines correctly when you have a busy day planned?". Higher levels of self-efficacy for medication-taking are reflected by higher scores in SEAMS

(continued)

Table 4.5   (continued)

Target behaviour	Determinant	Tool
Physical activity	Competence Autonomy Relatedness	*Basic Psychological Needs in Exercise Scale (BPNES)* (Vlachopoulos et al., 2010) The BPNES is a 12-item scale that measures the satisfaction of psychological needs for exercise; it comprises three constructs: Autonomy (items 3, 6, 9, 12), competence (items 1, 4, 7, 10) and relatedness (items 2, 5, 8, 11). Respondents indicate their degree of agreement with each statement on a five-point Likert scale ranging from 1 ("strongly disagree") to 5 ("strongly agree"). The maximum score for each construct is 20
	Motivation	*Exercise Motivations Inventory (EMI-2)* (Markland & Ingledew, 1997) A 51-item scale to measure motivation to exercise, including enjoyment, health pressures, social recognition and stress management. The EMI-2 scale encompasses 14 subscales, and each item is assessed using a 6-point Likert scale (0 = not at all true for me to 5 = very true for me). Example of an item: *Personally, I exercise (or might exercise) to have a healthy body*. Higher scores indicate higher exercise motivation
	Self-efficacy	*Self-Efficacy for Physical Activity (SEPA)* (Marcus et al., 1992) The SEPA scale assesses confidence for engaging in physical activity. By listing potential barriers, respondents have to indicate their confidence in a five-point Likert scale (1 = not confident to 5 = extremely confident). SEPA scale has five items
Diet	Different determinants (including motivation)	*Regulation of Eating Behaviour Scale (REBS)* (Pelletier et al., 2004) The REBS scale is a 24-item scale focusing on factors such as external regulation, identified regulation, introjected regulation, integrated regulation, amotivation and intrinsic motivation. By using a seven-point Likert scale (1 = does not correspond at all and 7 = corresponds exactly), respondents indicate the extent of motives for regulating their eating behaviour
Smoking cessation	Nicotine dependence	*The Fagerström Test for Nicotine Dependence* (Heatherton et al., 1991) The Fagerström test comprises six items to assess the quantity of cigarette consumption, the compulsion to smoke and dependence. It encompasses two types of responses: Yes/no (scored from 0 to 1) and multiple choice (scored from 0 to 3). The Fagerström score is obtained by summing the scores of individual items (min. 0 and max. 10). The higher the score is, the more intense is the person's physical dependence on nicotine

4.2 Tailoring Behaviour Change Techniques in the Development of an Intervention Plan

4.2.1 Behaviour Change Techniques to Support Chronic Disease Self-Management

To replicate and implement behaviour change interventions in practice, we need an agreed language to report their content. A reliable method has been developed to specify content in terms of behaviour change techniques (BCTs), the active components of a behaviour change intervention. A BCT is "an observable, replicable, and irreducible components of an intervention designed to alter or redirect causal processes that regulate behaviour" (Michie et al., 2013).

Based on a series of consensus exercises, an extensive hierarchically clustered taxonomy of 93 distinct BCTs has been developed (Michie et al., 2013) – BCT Taxonomy version 1 (BCTTv1). This taxonomy consists of a total of 16 clusters, covering a total of 93 BCTs, together with definitions and illustrative examples. BCTTv1 offers a reliable method for specifying, interpreting and implementing the active ingredients of interventions to change behaviours, which that can be helpful to professionals (Michie et al., 2013).

To facilitate access and support professional practice, a mobile application has been developed with a fully searchable version of BCTTv1 (https://www.ucl.ac.uk/behaviour-change/resources/online-tools-behaviour-change). BCTs can be searched by label or grouping or alphabetically.

BCTs, as active ingredients of the interventions, can take different functions such as education (i.e. increase knowledge or understanding), training (i.e. impart skills) or persuasion (i.e. use communication to induce or negative feelings to stimulate action). The most frequently used BCTs for education are information about health consequences (5.1), information about social and environmental consequences (5.3), feedback on behaviour (2.2), feedback on outcome(s) of behaviour (2.7) and self-monitoring of behaviour (2.3). Informing a person who smokes that the majority of people disapprove of smoking in public places is an example of using a BCT (information about social and environmental consequences 5.3) for education purposes. Explaining the likelihood of increasing the glycated haemoglobin levels (A1C) when adopting an unhealthy lifestyle is another example of a BCT used for education purposes (information about health consequences 5.1). Table 4.6 presents a set of BCTs with accompanying definition and examples (Michie et al., 2013).

For the self-management of chronic diseases, 21 core BCTs were identified from the BCTTv1 based on a literature search in conjunction with behavioural psychologists' feedback (Guerreiro et al., 2021). The 21 BCTs are common to the 5 target behaviours in the 7 high-priority chronic diseases considered in Chap. 3 (type 2 diabetes, COPD, obesity, heart failure, asthma, hypertension and ischaemic heart disease) and are available at Guerreiro et al. (2021). Additional BCTs were organised in supplementary sets per target behaviour (Guerreiro et al., 2021).

BCTs are designed to enable behaviour change and can do this by augmenting factors that facilitate behaviour change or by mitigating factors that inhibit

Table 4.6 Set of BCTs, accompanying definition and example of application

BCT	Definition	Example
1.1 Goal setting (behaviour)	Set or agree on a goal defined in terms of the behaviour to be achieved	Set the goal of eating five pieces of fruit per day as specified in public health guidelines
2.3 Self-monitoring of behaviour	Establish a method for the person to monitor and record their behaviour(s) as part of a behaviour change strategy	Give the person a pedometer and a form for recording the daily total number of steps
5.1 Information about health consequences	Provide information (e.g. written, verbal, visual) about health consequences of performing the behaviour	Explain that not finishing a course of antibiotics can increase susceptibility to future infection

From Michie et al. (2013) and Michie et al. (2014, p. 259, 262, 266)

behaviour change. An illustration of this point is the case of a person with type 2 diabetes who does not believe in her or his ability to increase physical activity. The BCT Graded tasks (8.7) – set easy-to-perform tasks, making them increasingly difficult, but achievable, until behaviour is performed (Michie et al., 2014) – might change the behaviour by increasing the belief about the person's capabilities. When promoting healthy eating, one might hypothesise that the BCT Restructuring the physical environment (12.1) – change or advise to change the physical environment in order to facilitate performance of the wanted behaviour or create barriers to the unwanted behaviour – might change this behaviour by eliminating the access to a vending machine with unhealthy snacks in the workplace.

Additional classifications of techniques to change behaviour and influence motivation have been developed. A notable example is the compendium of self-enactable techniques (Knittle et al., 2020), developed from existing taxonomies (e.g. BCTTv1, Kok et al., 2016). The compendium contains a list of 123 techniques that can be enacted by the individual, and each technique is presented with a label, a definition, instructive examples on health behaviours, its source, information on whether it requires external inputs (e.g. obtaining information) and prerequisite techniques (e.g. the technique "feedback on behaviour" can only be used if "self-monitoring of behaviour" is in place). This can be a valuable resource for intervention developers and recipients in the context of chronic disease management, e.g. in self-delivered and technology-assisted interventions.

There are benefits of using BCTs provided by a taxonomy in interventions to support behaviour change:

- To establish a structured link with behaviour determinants, which facilitates intervention tailoring and increases effectiveness.
- To specify intervention content, facilitating the identification of effective interventions in practice.
- To enhance the comprehensiveness of interventions in practice, as it is less likely that barriers and facilitators are disregarded when the intervention is tailored to behaviour determinants.
- To ensure consistency across interventions.

4.2.2 Tailoring Behaviour Change Techniques

Section 4.1, provides examples of behaviour determinants. As pointed out, tailoring the intervention to behaviour barriers increases the likelihood of success (Williams et al., 2019). For example, a pillbox or reminders will do little for a person deciding not to take a medication due to concerns about side effects; such barrier requires techniques increasing knowledge or understanding, such as information about health consequences (5.1), or inducing a feeling to stimulate action, such as pros and cons (9.2). These BCTs consist of, respectively, highlighting the positive and negative consequences of taking the medication and advising the person to compare reasons for wanting and not wanting to perform the behaviour (Michie et al., 2014).

As depicted in Fig. 4.1, tailoring BCTs can be seen as a two-step sequential process. Firstly, choosing BCTs that can potentially be used in the intervention, based on their alignment with behaviour barriers – Step 1 in Fig. 4.1. Secondly, selecting BCTs from this "list" and deciding on operationalisation according to the person's unique combination of e.g. morbidities, functional status, activities of daily living, preferences and resources – Step 2 in Fig. 4.1.

An important consideration is that it may be unnecessary and potentially inappropriate to deliver all BCTs listed in Step 1. As explained, the patient as a unique person should be considered when selecting a BCT addressing a behaviour barrier. For example, social support may not be suitable for a person living alone and having a restricted social network. Operationalising a selected BCT also requires attention to the patient as a unique person; for instance, advising a person to set reminders in a mobile phone to take the medication – prompts/cues (7.1) – may not be appropriate for older persons unfamiliar with these devices. In such a case, helpful alternatives may include using a post-it.

Another important consideration is about the use of the BCTs alone or in combination change. For example, for a person who forgets to take the medication, the BCT Prompts and cues (7.1) – introduce or define environmental or social stimulus with the purpose of prompting or cueing the behaviour – may be sufficient to overcome this barrier. A combination of BCTs may be needed for a person who has concerns about medication, such asusing information about health consequences (5.1) and pros and cons (9.2) in bundle. The two-step sequential process depicted in Fig. 4.1 aids the decision of suggesting BCTs alone or in combination.

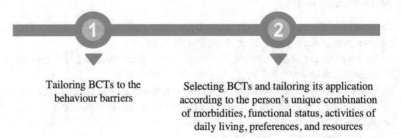

Fig. 4.1 Identifying and selecting BCTs when developing an intervention plan: steps 1 & 2

The alignment of BCTs with behaviour barriers is further exemplified in Table 4.7, using the case of physical activity of a fictitious person; the application of BCTs is also exemplified.

Table 4.7 Example: barriers to physical activity, aligned with BCTs and their application

COM-B component	Barrier	BCT	Definition (Michie et al., 2014)	Application
Physical capability	Fatigue	4.1 instruction on how to perform a behaviour	Advise or agree on how to perform the behaviour (includes "skills training")	Advise on physical activity or exercises that are less demanding
		1.4 action planning	Prompt detailed planning of performance of the behaviour (must incluvde at least one of context, frequency, duration and intensity). Context may be environmental (physical or social) or internal (physical, emotional or cognitive) (includes "implementation intentions")	Prompt the scheduling of physical activity for times in the day or week when the person feels less fatigue
Psychological capability	Lack of knowledge of the importance of physical activity	5.1 information about health consequences	Provide information (e.g. written, verbal, visual) about health consequences of performing the behaviour	Highlight the benefits for one's health of doing regular physical activity
Social opportunity	Lack of encouragement and support from family and friends	3.1 social support (unspecified)	Advise on, arrange or provide social support (e.g. from friends, relatives, colleagues, "buddies" or staff) or non-contingent praise or reward for performance of the behaviour. It includes encouragement and counselling, but only when it is directed at the behaviour	Advise the person to explain his/her interest in physical activity to friends and family and ask them to support his/her efforts
		3.2 social support (practical)	Advise on, arrange or provide practical help (e.g. from friends, relatives, colleagues, "buddies" or staff) for performance of the behaviour	Advise the person to invite friends and family to exercise with (e.g. walking in the park)

4.2.3 Selecting Behaviour Change Techniques According to the Length of the Intervention

The length of the intervention also influences the selection of BCTs. Box 4.3 presents the definition of brief and long-term interventions.

Box 4.3 Definition of Brief and Long-Term Intervention
Brief intervention

Intervention delivered in a short interaction between the provider and the individual, often carried out when the opportunity arises, typically taking no more than a few minutes. Although short in duration, a brief intervention can be delivered in several sessions (adapted from National Institute for Health and Care Excellence, 2014)

Long-term intervention

Intervention delivered in a longer interaction (e.g. around 30 minutes) between the provider and the individual, which has a structured plan and consists of multiple sessions over time (adapted from National Institute for Health and Care Excellence, 2014)

To facilitate comprehension, we have further conceptualised BCTs tailoring as a three-step sequential process, adding tailoring of BCTs to the length of the intervention as Step 3 (Fig. 4.2). That said, in practice Steps 2 and 3 can take place simultaneously.

In brief interventions it may not be feasible to use BCTs that require more than one encounter to operationalise. A good illustration is the case of Feedback on outcomes of behaviour (2.7). Let us consider a person living with obesity, who agrees to engage frequently in physical activity and is advised to monitor weigh (self-monitoring of outcome(s) of behaviour 2.4). It may be beneficial to provide feedback on how much weight the person has lost as an outcome of performing physical activity (Feedback on outcome(s) of behaviour 2.7). However, the selection of the

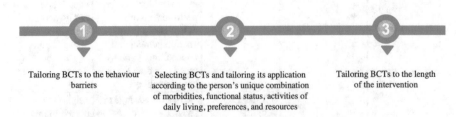

Tailoring BCTs to the behaviour barriers | Selecting BCTs and tailoring its application according to the person's unique combination of morbidities, functional status, activities of daily living, preferences, and resources | Tailoring BCTs to the length of the intervention

Fig. 4.2 Identifying and selecting BCTs when developing an intervention plan: steps 1, 2 & 3

latter BCT is not realistic in a brief intervention limited to one encounter or with unknown frequency of encounters. This can also be seen in the case of the BCT Review behaviour goal(s) (1.5), which may not be appropriate in a brief intervention limited to one encounter. However, if the brief intervention is delivered in several sessions, it is possible to use it provided that behaviour goals were previously set (Goal setting behaviour 1.1) and that is appropriate to revise them.

The number of BCTs used in an interaction may differ in a brief and long intervention. Due to the longer duration and structured nature, the latter may encompass a higher number of BCTs, if justified considering the behaviour determinants and person's unique preferences.

In brief and long interventions, professionals should also consider the modes of delivery of BCTs. Modes of delivery (MoD) are defined as the method(s) by which the content (i.e. BCTs) is brought to the person (Marques et al., 2020); they can influence the effectiveness of the interventions. For example, meta-research found effectiveness to be higher in smoking cessation interventions when the BCTs were delivered in person as opposed to written form (Black et al., 2020).

The modes of delivery are organised in four-level hierarchical structures comprising 65 entities. There are 15 upper-level classes, such as Informational MoD and Environmental change MoD. Each upper-level class includes sub-classes, as represented in Table 4.8 for Informational MoD.

Many factors influence the decision on the modes of delivery, not only the length of the interventions, but also the preferences and needs of the person. The modes of delivery should be considered when developing an intervention plan.

Key Points
- A plethora of determinants can influence positively or negatively the key self-management behaviours in high-priority chronic diseases.
- The COM-B model can guide the identification of behaviour determinants.
- Specific tools and approaches can be useful for assessing behaviour determinants such as Beliefs About Medicines Questionnaire (BMQ), Regulation of Eating Behaviour Scale (REBS) and interview.
- Behaviour change techniques (BCTs) are the active components of behaviour change interventions.
- When developing an intervention plan, tailoring BCTs should consider the behaviour determinants identified for the person, her/his unique combination of morbidities, functional status, activities of daily living, preferences, resources and context, and the length of the intervention.
- When behaviour change techniques are implemented in practice, consideration should also be given to the mode of delivery.

Table 4.8 Example of modes of delivery including in the Informational MoD

Upper-level class	Sub-level 1	Sub-level 2	Definition	Examples of usage
Informational MoD			Mode of delivery that involves intentional transmission of a representation of the world to an intervention recipient with the aim of changing that person's representation of the world	This includes delivery of rewards, prompts and cues that result in learning and information about the environment and environmental contingencies
	Human interactional mode of delivery		Informational mode of delivery that involves a person as intervention source who interacts with an intervention recipient in real time	
		Face-to-face mode of delivery	Human interactional mode of delivery that involves an intervention source and recipient being together in the same location and communicating directly	
		At-a-distance mode of delivery	Human interactional mode of delivery that involves an intervention source and recipient being in different locations and communicating through a communication channel	
	Printed material mode of delivery		Informational mode of delivery that involves use of printed material	Can include paper, acetate, text, diagrams and photographic images
		Letter mode of delivery	Printed material mode of delivery that involves a letter or postcard that can be sent through the post or handed directly to the recipient	
		Printed publication mode of delivery	Printed material mode of delivery that involves use of a printed publication	Includes leaflets, brochures, newspapers, newsletter, booklets, magazines, manuals or worksheets
	Electronic mode of delivery		Informational mode of delivery that involves electronic technology in the presentation of information to an intervention recipient	
		Mobile digital device mode of delivery	Electronic mode of delivery that involves presentation of information by a handheld mobile digital device that can store, retrieve and process data	

From Marques et al. (2020, p. 9, 10).

References

Black, N., Eisma, M. C., Viechtbauer, W., Johnston, M., West, R., Hartmann-Boyce, J., et al. (2020). Variability and effectiveness of comparator group interventions in smoking cessation trials: a systematic review and meta-analysis. *Addiction, 115*(9), 1607–1617. https://doi.org/10.1111/add.14969

Brown, T. J., Twigg, M., Taylor, N., Easthall, C., Hartt, J., Budd, T., … Bhattacharya, D. (2017). Final report for the IMAB-Q study: validation and feasibility testing of a novel questionnaire to identify barriers to medication adherence.

Chean, K. Y., Goh, L. G., Liew, K. W., Tan, C. C., Choi, X. L., Tan, K. C., & Ooi, S. T. (2019). Barriers to smoking cessation: a qualitative study from the perspective of primary care in Malaysia. *BMJ Open, 9*(7), 1–8. https://doi.org/10.1136/bmjopen-2018-025491

Cheng, L. J., Wu, V. X., Dawkes, S., Lim, S. T., & Wang, W. (2019). Factors influencing diet barriers among outpatients with poorly-controlled type 2 diabetes: a descriptive correlational study. *Nurs. Health Sci., 21*(1), 102–111. https://doi.org/10.1111/nhs.12569

Cortis, C., Puggina, A., Pesce, C., Aleksovska, K., Buck, C., Burns, C., et al. (2017). Psychological determinants of physical activity across the life course: A "DEterminants of DIet and Physical ACtivity" (DEDIPAC) umbrella systematic literature review. *PLoS One, 12*(8), 1–25. https://doi.org/10.1371/journal.pone.0182709

Cradock, K. A., Quinlan, L. R., Finucane, F. M., Gainforth, H. L., Martin Ginis, K. A., de Barros, A. C., et al. (2021). Identifying barriers and facilitators to diet and physical activity behaviour change in type 2 diabetes using a design probe methodology. *J. Personalized Med., 11*(2), 1–26. https://doi.org/10.3390/jpm11020072

Félix, I. B., & Henriques, A. (2021). Medication adherence and related determinants in older people with multimorbidity: a cross-sectional study. *Nurs. Forum, 1–10.* https://doi.org/10.1111/nuf.12619

Ferra, J. R. M., Vieira, A. C., Carvalho, J. S., Matos, C., & Nogueira, F. (2019). Barriers to smoking cessation: the patient's perspective. *Eur. Respir. J., 54,* PA2852. https://doi.org/10.1183/13993003.congress-2019.pa2852

Guerreiro, M. P., Strawbridge, J., Cavaco, A. M., Félix, I. B., Marques, M. M., & Cadogan, C. (2021). Development of a European competency framework for health and other professionals to support behaviour change in persons self-managing chronic disease. *BMC Med. Edu., 21*(1), 1–14. https://doi.org/10.1186/s12909-021-02720-w

Hankonen, N., & Hardeman, W. (2020). Developing behavior change interventions. In M. S. Hagger, L. D. Cameron, K. Hamilton, N. Hankonen, & T. Lintunen (Eds.), *The handbook of behavior change* (pp. 300–317). Cambridge University Press.

Heatherton, T., Kozlowski, L., Frecker, R., & Fagerstrom, K.-O.-O. (1991). The Fagerström test for nicotine dependence: a revision of the Fagerstrom Tolerance Questionnaire. *Br. J. Addict., 86*(9), 1119–1127. https://doi.org/10.1111/j.1360-0443.1991.tb01879.x

Horne, R., Weinman, J., & Hankins, M. (1999). The beliefs about medicines questionnaire: the development and evaluation of a new method for assessing the cognitive representation of medication. *Psychol. Health, 14*(1), 1–24. https://doi.org/10.1080/08870449908407311

Jackson, C., Eliasson, L., Barber, N., & Weinman, J. (2014). Applying COM-B to medication adherence: a suggested framework for research and interventions. *Eur. Health Psychol. Soc., 16*(1), 7–17.

Kardas, P., Lewek, P., & Matyjaszczyk, M. (2013). Determinants of patient adherence: a review of systematic reviews. *Front. Pharmacol., 4*(91), 1–16. https://doi.org/10.3389/fphar.2013.00091

Kelly, S., Olanrewaju, O., Cowan, A., Brayne, C., & Lafortune, L. (2018). Alcohol and older people: A systematic review of barriers, facilitators and context of drinking in older people and implications for intervention design. *PLoS One, 13*(1), 1–14. https://doi.org/10.1371/journal.pone.0191189

Keyworth, C., Epton, T., Goldthorpe, J., Calam, R., & Armitage, C. J. (2020). Acceptability, reliability, and validity of a brief measure of capabilities, opportunities, and motivations ("COM-B"). *Br. J. Health Psychol., 25*(3), 474–501. https://doi.org/10.1111/bjhp.12417

Knittle, K., Heino, M., Marques, M. M., Stenius, M., Beattie, M., Ehbrecht, F., et al. (2020). The compendium of self-enactable techniques to change and self-manage motivation and behaviour v.1.0. *Nat. Human Behaviour, 4*(2), 215–223. https://doi.org/10.1038/s41562-019-0798-9

Kok, G., Gottlieb, N. H., Peters, G. J., Mullen, P. D., Parcel, G. S., Ruiter, R. A., Fernández, M. E., Markham, C., & Bartholomew, L. K. (2016). A taxonomy of behaviour change methods: an Intervention Mapping approach. *Health Psychol. Rev., 10*(3), 297–312. https://doi.org/10.1080/17437199.2015.1077155

Kvarnström, K., Westerholm, A., Airaksinen, M., & Liira, H. (2021). Factors contributing to medication adherence in patients with a chronic condition: a scoping review of qualitative research. *Pharmaceutics, 13*(7), 1–41. https://doi.org/10.3390/pharmaceutics13071100

Marcus, B. H., Selby, V. C., Niaura, R. S., & Rossi, J. S. (1992). Self-efficacy and the stages of exercise behavior change. *Res. Q. Exerc. Sport, 63*(1), 60–66. https://doi.org/10.1080/02701367.1992.10607557

Markland, D., & Ingledew, D. K. (1997). The measurement of exercise motives: factorial validity and invariance across gender of a revised exercise motivations inventory. *Br. J. Health Psychol., 2*, 361–376. https://doi.org/10.1111/j.2044-8287.1997.tb00549.x

Marques, M. M., Carey, R. N., Norris, E., Evans, F., Finnerty, A. N., Hastings, J., et al. (2020). Delivering behaviour change interventions: development of a mode of delivery ontology. *Wellcome Open Res., 5*(125). https://doi.org/10.12688/wellcomeopenres.15906.1

Michie, S., van Stralen, M. M., & West, R. (2011). The behaviour change wheel: a new method for characterising and designing behaviour change interventions. *Implement. Sci., 6*(42), 1–11. https://doi.org/10.1186/1748-5908-6-42

Michie, S., Richardson, M., Johnston, M., Abraham, C., Francis, J., Hardeman, W., et al. (2013). The behavior change technique taxonomy (v1) of 93 hierarchically clustered techniques: Building an international consensus for the reporting of behavior change interventions. *Ann. Behav. Med., 46*, 81–95. https://doi.org/10.1007/s12160-013-9486-6

Michie, S., Atkins, L., & West, R. (2014). *The behavior change wheel: a guide to designing interventions (First edit)*. Silverback Publishing.

Mishra, P., Vamadevan, A. S., Roy, A., Bhatia, R., Naik, N., Singh, S., et al. (2021). Exploring barriers to medication adherence using com-b model of behaviour among patients with cardiovascular diseases in low-and middle-income countries: a qualitative study. *Patient Prefer. Adherence, 15*(June), 1359–1371. https://doi.org/10.2147/PPA.S285442

National Institute for Health and Care Excellence. (2014). NICE Guidance: behaviour change: individual approaches. Retrieved from https://www.nice.org.uk/guidance/ph49

Pelletier, L. G., Dion, S. C., Slovinec-D'Angelo, M., & Reid, R. (2004). Why do you regulate what you eat? Relationships between forms of regulation, eating behaviors, sustained dietary behavior change, and psychological adjustment. *Motiv. Emot., 28*(3), 245–277. https://doi.org/10.1023/B:MOEM.0000040154.40922.14

Pinho, M. G. M., Mackenbach, J. D., Charreire, H., Oppert, J. M., Bárdos, H., Glonti, K., et al. (2018). Exploring the relationship between perceived barriers to healthy eating and dietary behaviours in European adults. *Eur. J. Nutr., 57*(5), 1761–1770. https://doi.org/10.1007/s00394-017-1458-3

Risser, J., Jacobson, T., & Kripalani, S. (2007). Development and psychometric evaluation of the self-efficacy for appropriate medication use scale (SEAMS) in low-literacy patients with chronic disease. *J. Nurs. Meas., 15*(3), 203–219. https://doi.org/10.1891/106137407783095757

Ritter, P. L., Lorig, K., & Laurent, D. D. (2016). Characteristics of the Spanish- and English-language self-efficacy to manage diabetes scales. *Diabetes Edu., 42*(2), 167–177. https://doi.org/10.1177/0145721716628648

Seguin, R., Connor, L., Nelson, M., Lacroix, A., & Eldridge, G. (2014). Understanding barriers and facilitators to healthy eating and active living in rural communities. *J. Nutr. Metabolism, 2014*, 23–25. https://doi.org/10.1155/2014/146502

Vlachopoulos, S. P., Ntoumanis, N., & Smith, A. L. (2010). The basic psychological needs in exercise scale: translation and evidence for cross-cultural validity. *Int. J. Sport Exercise Psychol., 8*(4), 394–412. https://doi.org/10.1080/1612197X.2010.9671960

West, R. & Michie, S. (2020). A brief introduction to the COM-B Model of behaviour and the PRIME Theory of motivation. *Qeios*. https://doi.org/10.32388/WW04E6.

Williams, D. M., Rhodes, R. E., & Conner, M. T. (2019). Conceptualizing and intervening on affective determinants of health behaviour. *Psychol. Health, 34*(11), 1267–1281. https://doi.org/10.1080/08870446.2019.1675659

Open Access This chapter is licensed under the terms of the Creative Commons Attribution 4.0 International License (http://creativecommons.org/licenses/by/4.0/), which permits use, sharing, adaptation, distribution and reproduction in any medium or format, as long as you give appropriate credit to the original author(s) and the source, provide a link to the Creative Commons license and indicate if changes were made.

The images or other third party material in this chapter are included in the chapter's Creative Commons license, unless indicated otherwise in a credit line to the material. If material is not included in the chapter's Creative Commons license and your intended use is not permitted by statutory regulation or exceeds the permitted use, you will need to obtain permission directly from the copyright holder.

Chapter 5
Communication and Person-Centred Behaviour Change

Afonso Miguel Cavaco, Carlos Filipe Quitério, Isa Brito Félix, and Mara Pereira Guerreiro

Learning Outcomes

This chapter contributes to achieving the following learning outcomes:

BC6.1 Generate with the person opportunities for behaviour change.

BC6.2 Assess the extent to which the person wishes and can become co-manager of his/her chronic disease.

BC6.3 Demonstrate how to promote coping skills and self-efficacy to manage chronic disease's physical, emotional and social impacts in everyday life.

BC6.4 Assist the person to become co-manager of his/her chronic disease in partnership with health professionals.

BC7.1 Apply strategies to support the cooperative working relationship between the person and a healthcare provider.

BC7.2 Demonstrate active listening of the person's concerns and difficulties in the self-management of chronic disease.

BC14.1 Share information and adequate educational materials according to individual factors (e.g. knowledge gaps, health literacy level and preferences).

A. M. Cavaco (✉)
Faculty of Pharmacy University of Lisbon, Lisbon, Portugal
e-mail: acavaco@ff.ulisboa.pt

C. F. Quitério
Centro Hospitalar de Setúbal, Setúbal, Portugal

I. B. Félix
Nursing Research, Innovation and Development Centre of Lisbon (CIDNUR),
Nursing School of Lisbon, Lisbon, Portugal

M. P. Guerreiro
Nursing Research, Innovation and Development Centre of Lisbon (CIDNUR),
Nursing School of Lisbon, Lisbon, Portugal

Egas Moniz Interdisciplinary Research Center (CiiEM), Egas Moniz School of Health
& Science, Monte de Caparica, Portugal

© The Author(s) 2023
M. P. Guerreiro et al. (eds.), *A Practical Guide on Behaviour Change Support
for Self-Managing Chronic Disease*, https://doi.org/10.1007/978-3-031-20010-6_5

5.1 Overview of Key Concepts

Supporting successful behaviour change interventions requires relational and communication skills. It is well accepted that communication adopted by professionals can foster engagement in behaviour change or, if suboptimal, bears a detrimental effect.

Concepts such as patient-centred communication and shared decision-making are pivotal in behaviour change interventions; these two concepts remind health and other professionals that they must embrace the idea that change happens within each person, not through professionals' willingness.

In this chapter, we will address the essential features of a relationship between the professional and the person when supporting the change or the maintenance of a self-management behaviour.

5.1.1 Patient Empowerment

A current conceptualisation of patient empowerment posits that it occurs when patients make autonomous, informed decisions about their health, supported by a professional, to increase their capacity to think critically and make independent and informed decisions (Anderson & Funnell, 2010).

One definition of patient empowerment considers that patients are empowered when they have the knowledge, skills, attitudes and self-awareness necessary to influence their and others' behaviours to improve the quality of their lives (Funnell et al., 1991).

Another definition defines patient empowerment in the healthcare context as promoting autonomous self-regulation to maximise the individual's potential for health and wellness (Lau, 2002). As one can realise, patient empowerment begins with information and education and includes seeking out information about one's condition and actively participating in treatment decisions.

How does it start? Patient empowerment begins when professionals acknowledge that patients with chronic diseases are in control of their daily care. Professionals must recognise that the most significant impact on a person's health and well-being results from their own management decisions and daily actions.

The professionals involved in patient empowerment make clear to patients that being in control of their daily self-management decisions comes with responsibility for those decisions and the resulting consequences. Responsibility, in turn, means that patients cannot surrender their control over chronic disease self-management, no matter how much they wish.

While professionals cannot control and therefore cannot be responsible for the self-care decisions, they are accountable for ensuring that their patients make informed self-management decisions. Here, "informed" means an adequate understanding of self-management and an awareness of the aspects of their personal lives that influence self-management decisions.

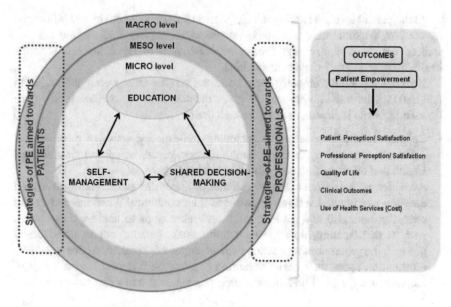

Fig. 5.1 The EMPATHiE conceptual framework. (Adapted from Kayser et al., (2019) and European Patients' Forum (2017))

A conceptual framework proposed by the EMPATHiE project considers patient empowerment on three different levels: macro, meso and micro, as depicted in Fig. 5.1. All associated variables work as moderators of patient empowerment (Kayser et al., 2019).

- Macro-level, i.e. the vision circulated by the authorities. This level comprises the definition of the joint plans at the organisational level, e.g. shorter hospital stays.
- Meso-level, i.e. the medicine level and the perspective of patients living with multimorbidity on professionals and informal caregivers' role. It comprises patient-centred care and autonomy support.
- Micro-level, i.e. patients' perspective stemming from their personal health experiences. This level comprises patient participation through shared decision-making to foster involvement.

In this chapter, the focus will be on the meso- and micro-level patient empowerment. Empowering patients encompasses concepts such as patient-centredness, shared decision-making, motivational interviewing, counselling and signposting to support services. According to Holmström & Röing (2010), patient-centeredness as a process is of great value in patient empowerment. Patient empowerment can be achieved by patient-centeredness, but patients can also empower themselves. Nevertheless, one should not forget that patient empowerment is influenced by attributes of professionals, such as individual features, training, personal values and professional goals.

At the patient level, the ability to engage in empowerment activities is influenced by context, personal characteristics, values, social support and disease circumstances (e.g. duration, severity). Professionals also need to pay attention to patient's health literacy and self-management skills.

How can a patient's empowerment be assessed? The empowered patient has to feel like they have the ability and are given the confidence to manage their condition. Patients can indicate this state through their:

- Capacities, beliefs or resources including self-efficacy, sense of meaning and coherence about their condition, health literacy, perceived control and feelings. Health and other professionals should respect all these.
- Activities or behaviours (things patients do), such as participating in shared decision-making and self-management of their condition. When the patient can choose meaningful and realistic goals and takes steps to achieve those goals, such as participating in collective activities (e.g. patient support or advocacy groups), the professional is dealing with an empowered patient. Active search for information about their health condition (e.g. on the Internet) is also a sign of empowerment, even if misinformation is present, which is a distinct issue.

5.1.2 Person-Centredness

This section deals with another well-known concept in healthcare provision, known as patient-centredness.

The terms person-centred, people-centred and patient-centred will be used interchangeably. Looking at the Mead & Bower (2002) definition of patient-centred care, it is the understanding of the patient as a unique human being or entering that person's world to see illness through their perspective.

Person-centredness supports the care of the whole person (negative and positive aspects), for the person (assisting the fulfilment of life projects), by the person (competent and high ethical conduct) and with the person (respectful collaboration). However, the concept can be seen as further complex. According to a systematic review from Scholl et al. (2014), patient-centredness encompasses 15 dimensions, organised in principles, enablers and activities (see Fig. 5.2):

- Principles: essential characteristics of professionals, professional-patient relationship, patient as a unique person, the biopsychosocial perspective.
- Enablers: integration of medical and non-medical care, teamwork and team building, access to care, coordination and continuity of care.
- Activities: patient information, patient involvement in care, involvement of family and friends, patient empowerment, physical support, emotional support.

One can realise from all previous dimensions that some are more dependable on the healthcare professional, e.g. professional-patient communication and relationship and patient involvement in care and empowerment, than others, e.g. access to care and medical and non-medical care integration.

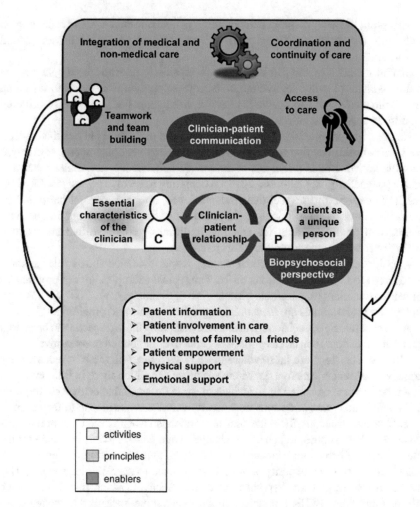

Fig. 5.2 A model of person-centeredness. (Scholl et al., 2014)

Although robust evidence about the positive effects of person-centredness in health endpoints is still lacking, review studies have provided objective and positive accounts, such as the reduction of inappropriate prescription and use of benzodiazepines and related drugs (Mokhar et al., 2018), as well as improvements in the clinical outcomes of persons with type 2 diabetes in primary healthcare (Vuohijoki et al., 2020).

5.1.3 Shared Decision-Making

When making health-related decisions, some people prefer those decisions to be taken and controlled by healthcare professionals. Other persons are willing to participate in decision-making and enjoy a degree of autonomy when managing their conditions.

Assessing the level of direct responsibility in decision-making, and reflecting on a person's empowerment, is an area of research and education named shared decision-making (SDM).

Shared decision-making in healthcare is especially relevant for the success of health behaviour change interventions. It requires applying strategies to create an environment conducive to open and effective communication, from active listening to building a trustful relationship.

Reasons for incorporating the person's views and preferences in healthcare decisions are twofold. The first reason is an ethical imperative. This imperative emerges from the known ethical principles of autonomy, beneficence, non-maleficence and justice (Beauchamp & Childress, 2019). Attempting to involve persons in decision-making is expected to respect their independence, do good, prevent harm and be equitable. Ensuring these principles in practice and involving persons in decision-making can be challenging; for example, cultural or cognitive barriers make it harder for them to understand the information.

Another reason, from a practical nature, is that evidence shows that persons' preferences for healthcare options differ. The typical examples are the preferences for cancer treatment (e.g. Hamelinck et al., 2014; Stalmeier et al., 2007); it has been shown that professionals are inaccurate in predicting persons' preferences.

At least in principle, decisions in healthcare (including those respecting to behaviour change interventions) can be categorised as "effective" or "preference-sensitive".

Effective decisions are those where there is agreement on the best management strategy. Preference-sensitive decisions are those for which there is little evidence on the best course of action or for which, despite the available evidence, weighing the benefits and harms of options may greatly vary from one person to the other.

In treatment decisions, SDM has been recommended when a decision is preference-sensitive. When it comes to behaviour change, while there is robust evidence for the effectiveness of behaviour change techniques (BCTs) in promoting sustained health behaviour changes, the majority of the existing research tests BCTs in groupings (i.e. bundles of BCTs that are hypothesised to be effective), which means that when selecting individual BCTs in practice, professionals sometimes need to make preference-sensitive decisions. There is, nonetheless, a growing body of research using optimisation designs to test individual BCTs and compare their effectiveness. SDM is, therefore, a means to ensure that people's views are incorporated in the process, leading, hopefully, to decisions that fit their beliefs and preferences. The general effects of SDM mainly relate to outcomes on the cognitive level (e.g. showing command of the treatment options available and their implications) and affective level (e.g. showing acceptance or denial of a treatment option), especially when persons perceive to have been involved in decisions (Agbadjé et al., 2020).

The literature offers examples of SDM shared decision in behaviour change in chronic disease, such as approaches to initiate behaviour change in persons with cardiovascular disease (Cupples et al., 2018).

Some authors devoted attention to the intervention functions and behaviour change techniques (BCTs) needed to implement SDM in clinical practice, i.e. BCTs to achieve SDM, a different focus from employing BCTs as part of an intervention plan to achieve self-management behaviours. Intervention functions are "a broad category of means by which an intervention can change behaviour" (Michie et al., 2011). For instance, the intervention function "education", alone or in combination with other intervention functions (e.g. education + persuasion, education + training + modelling + enablement), was associated with effective SDM implementation. Examples of BCTs associated with effective SDM implementation were instruction on how to perform the behaviour, demonstration of the behaviour, feedback on behaviour, pharmacological support, material reward and biofeedback (Agbadjé et al., 2020).

One of the most cited approaches for achieving SDM is the work from Elwyn et al., (2012), updated by the same authors in 2017 (Elwyn & Durand, 2017). SDM depends on activities that help confer agency, which refers to the capacity of individuals to act independently and make their own free choices.

SDM aims to confer agency by two activities:

1. Providing high-quality information
    Based on knowledge acquired previously and during the intervention, the person can assess what is important concerning the outcomes associated with different options, processes and paths that lead to these outcomes.

2. Supporting deliberation
    Support the person to deliberate about their options by exploring their reactions to the information provided. When offered a role in decisions, persons can be surprised, unsettled by the possibilities and uncertain about what might be best.

For instance, in the field of behaviour change in health promotion, interventions to increase levels of SDM take two forms (Gültzow et al., 2021):

*   Training healthcare professionals (and/or persons) in BCTs.
*   Decision aids to be used before, during or following consultations or more generic question prompt lists. Decision aids can potentially be beneficial in supporting people to change preventive health behaviours, especially regarding smoking.

5.1.4   Health Literacy and Education

The linkage between behaviour change and health literacy is well recognised. Self-management behaviours may be promoted when enhancing a person's health knowledge through education, even if the relationship between education and behaviour change is not linear (Walters et al., 2020).

One note should be added regarding the difference between information and education. While information is predominantly one-way communication, from the professional to the person, education aims to confirm the person has acquired the knowledge and makes excellent and independent use of it.

Health literacy can be defined as the achievement of a level of knowledge, personal skills and confidence to take action to improve individual and community health by changing one's lifestyle and living conditions. A useful resource for this and other definitions is the World Health Organization (WHO) Health Promotion Glossary, commissioned in 1986 and updated in 2021 (Nutbeam & Muscat, 2021).

Thus, health literacy means more than reading pamphlets and making appointments (i.e. functional literacy). It involves improving persons' access to health information and the role of professionals in developing their capacity to use it effectively (i.e. interactive and critical literacy).

Professionals in brief or long interventions can contribute to improving a person's health literacy through generic education and by using BCTs that increase knowledge, understanding or impart skills (training).

Several approaches can be used to provide tailored education about a condition its treatment or self-management behaviours; a simple approach is the "Elicit-Provide-Elicit" technique, described below. The person should do most of the talking; this allows an understanding of the person's current knowledge and potential knowledge gaps and misconceptions. Personal views must be heard and listened to, even if the professional thinks they are incorrect (Bull & Dale, 2021).

First, the professional finds out what a person already knows about a condition, treatment or self-management behaviour, plus what she or he would like to know ("Elicit"), by posing questions such as "What do you know about X?". Then, the professional asks for permission and provides information that is helpful for the person ("Provide"), in a non-judgemental way, e.g. "Others have benefited from", "What we know about X is". Next, the professional checks the person's understanding, interpretation or response to what has been said ("Elicit") e.g. - "What do you think about this information I gave you?" "What questions do you have?".

A critical success factor in providing education or training is how the professional communicates. Fear tactics have no place in person-centred behaviour change communication. It is also important to remember that some people need time to adjust to new information, master new skills or make short- or long-term behaviour changes.

Simple rules to maximise effective education are:

- Using preferably tangible support, either printed or digital, depending on the person's preference, and factors such as literacy, numeracy and culture (e.g. brochures, podcasts, YouTube videos, videos, PowerPoint presentations, posters or charts or models).
- Combining text with graphics and pictures, instead of long written instructions only.
- Asking open-ended questions to assess the person's understanding of printed or online materials.
- Speaking at a moderate pace, especially when providing instructions.
- Respecting the person's limits, offering only the amount of information that an individual can handle at one time.
- Using plain language, avoiding complicated medical terminology or jargon to minimise the risk of misunderstanding.

Box 5.1 Examples of Simpler Wording

"Swallow" instead of "take" the medicine.
"Harmful" instead of "adverse" effect.
"Fats" instead of "lipids" reduction in daily food.
"Belly" instead of "abdomen" perimeter as an indicator of cardiovascular risk.
"Lasting a short time, but often causing a serious problem" instead of "acute"
    conditions.

The type of educational resources that a person responds to varies from person to person. Using a mixed media approach often works best. Professionals must review these resources before sharing them with persons living with chronic diseases. When developing their own materials, professionals should seek some form of validation, e.g. by pretesting the materials in a controlled sample and evaluating knowledge and/or skills acquisition.

Table 5.1 presents educational materials to support behaviour change interventions from reliable sources. Other online resources include webpages such as the WHO, national centres for disease control (e.g. health directorates) and other national health organisations (e.g. medicines agencies).

There is a constantly growing number of digital resources, such as websites and mobile applications. Many have commercial purposes, variable information quality and trustworthiness. One tool to aid the choice of digital health apps is ORCHA (https://orchahealth.com/services/digital-health-libraries/). HON (https://www.hon.ch/en/) certifies quality health information on websites.

Table 5.1 Examples of educational materials available as webpages

Target behaviour	Resources
Diet (including alcohol intake)	https://www.nhs.uk/live-well/healthy-weight/ start-the-nhs-weight-loss-plan/ https://www.bda.uk.com/food-health/your-health/obesity-and-overweight.html https://www.bda.uk.com/food-health/food-facts.html https://www.nhs.uk/live-well/eat-well/
Physical activity	https://www.nhs.uk/live-well/exercise/ why-sitting-too-much-is-bad-for-us/
Medication adherence	https://www.fda.gov/consumers/consumer-updates/ are-you-taking-medication-prescribed
Smoking cessation	https://www.nhs.uk/live-well/quit-smoking/ https://www.helpguide.org/articles/addictions/how-to-quit-smoking.htm#
Symptom monitoring and management	https://www.helpguide.org/articles/healthy-living/blood-pressure-and-your-brain.htm

5.2   Providing Person-Centred Behaviour Change Support

Effective communication strategies are paramount for collaborative planning, understanding the person's situation and successfully supporting behaviour change, either in single, opportunistic or repeated and more extended interactions.

As explained in Chap. 4, brief interventions are delivered in a short interaction between the provider and the person, often carried out when the opportunity arises, typically taking no more than a few minutes. Although shorter, a brief intervention can be delivered in several sessions. Brief interventions are often the only practical way of supporting behaviour change.

Long-term interventions are delivered in extended interactions (e.g. around 30 minutes) between the professional and the person, following a structured plan and multiple sessions over time.

Either brief or long, behaviour change interventions tend to be nested in existing encounters. Both interventions involve plans developed in cooperation with the person. One well-accepted approach to co-generating collaborative plans is using person-centred communication and mechanisms to involve the person in decisions; these two topics will be explored in the next sections.

5.2.1   Basic Communication Skills

Effective behaviour change interventions require two basic communication skills: good questioning and adequate listening.

Questions can be divided into four main types, detailed below.

1. **Open-ended questions,** which are questions that cannot be answered with a simple yes, no or another preconceived response. Often, they start with "What", "How" or "How come". They facilitate communication by encouraging the person to describe or explain the health or behaviour issue in their own words. Examples of these questions include "Please, tell me more about your smoking behaviour" and "What are the triggers that make you crave for a cigarette?".

   Open-ended questions are frequently used early in the interaction process, to expand the dialogue and encourage the person to tell his or her story.

   An interaction with a predominance of open-ended questions will be lengthier, and there is a possibility of handling less relevant information.

2. **Focused or closed-ended questions,** which are questions that direct the person to a specific answer. These questions can frequently be answered by yes, no or through a simple, definitive reply, as exemplified by "Do you always smoke after meals?"

Focused questions are frequently used to fine-tune the understanding of a problem and clarify details that the person may not have addressed in response to an open-ended question.

If the preceding questions have been open, closed questions may result in a more detailed answer than just plain and straightforward facts, as shown in Box 5.2.

Box 5.2 Example of Detailed Information Collected with a Focused Closed Question Proceeded by Open Questions

Professional: "Do you always smoke after meals?"
Person: "Yes, it gives more 'me' time".

An interaction with mostly closed-ended questions will limit the person's participation and increase the risk of collecting a different story from what the person truly experiences.

3. **Circular questions** are a more sophisticated information-gathering technique that asks the person to provide information from a different or someone else's perspective (e.g. "How might this problem change in the near future?"). These questions are often very effective for gaining an understanding of the subtle nuances of a problem.

4. **Leading or directive questions** suggest a correct answer, as evident in the case of "You know that smoking is bad for your health, right?". These questions introduce undue bias in a person's accounts and should be avoided.

Listening skills are crucial to demonstrate that the person has been heard and understood. Listening involves responses to the content, feelings or both. There are four general types of responses to the content presented next.

1. **Nodding** means providing non-verbal feedback by using paralanguage and head movements to encourage the person to talk, as illustrated in Box 5.3.

Box 5.3 Example of Nodding

Person: I have had issues with quitting smoking… for several years. I have tried… the last time was…
Professional (looks into the person, nods his head): hum-hum… (i.e., I am listening).

2. **Parroting** consists of repeating the last few words that the person said, and it is the most straightforward responding skill. This response demonstrates that the professional is listening and frequently encourages the person to continue and/or elaborate, as exemplified in Box 5.4.

Box 5.4 Example of Parroting

Person: This last month, I felt that I had to make another attempt to quit smoking…
Professional: Quit smoking…

3. **Paraphrasing** is slightly more sophisticated than parroting. It offers key points for a small amount of content and helps check the accuracy of what the professional has heard (Box 5.5). This response reveals higher attention levels. It provides the possibility to correct any misconceptions and demonstrates attention.

Box 5.5 Example of Paraphrasing

Person: Well, I started smoking 15 years ago. I tried several times to quit smoking and even managed to be two months without smoking. I have heard about consultations out there, but I want none of that. It is about my will…to stop smoking.
Professional: So, I understand you want to quit smoking. What makes you crave a cigarette?

4. **Summaries** are another form of responding to content. They are usually lengthier and deal with a more significant amount of information than paraphrasing or parroting (Box 5.6). Summaries are used throughout an interaction to make sure the narrative is heard wholly and correctly. Its main objective is to explore the content of the dialogue or bring it to a conclusion.

Box 5.6 Example of a Summary

Person: I get up from my desk to take a break, and I suddenly feel that I really want to light up. I feel a strong urge to smoke in all my breaks at work. It really helps me to unwind.
Professional: You are telling me that smoking when you have a break at work helps you relax, so that is a trigger for smoking in your case…

Paraphrasing and summaries are examples of **active listening**. The listener highlights the main ideas expressed by the person, with summarising allowing for further content development in the next communication cycle, by reviewing in more detail a previous one.

Active listening skills also comprise **reflective responses,** in which the listener identifies feelings and emotions. Responses to feelings can take different forms and are particularly useful for demonstrating proper understanding and empathy.

There are three types of reflective responses:

1. **Sympathetic responses** communicate how the professional feels about what has been said or happened, centred on the professional's own framework (Box 5.7). It does not consistently demonstrate an emotional synchronisation with the person's feelings.

Box 5.7 Example of a Sympathetic Response

Person: Well, I started smoking 15 years ago. I tried several times to quit smoking and even managed to go two months without smoking. I know consultations are a good option, but I would like to try on my own. It is about my willpower... to stop smoking.
Professional: Professionals can offer help to quit smoking.

---

2. **Empathetic responses** demonstrate that the professional understands and accepts the feelings that the person has experienced, centred on the person's agenda (Box 5.8). To be effective, it must be perceived as genuine and sincere. This emotional resonance is essential to build acceptance of behaviour change support.

Box 5.8 Example of an Empathetic Response

Person: Well, I started smoking 15 years ago. I tried several times to quit smoking and even managed to go two months without smoking. I have heard about consultations out there, but I want none of that. It is about my will...to stop smoking.
Professional: It is hard to quit smoking after several years, and trying certainly requires motivation. Would you like to talk about smoking cessation?

---

3. **Denial responses** contradict what the person expresses and undermine the interview process (Box 5.9). Like leading questions, denial responses should be avoided.

Box 5.9 Example of a Denial Response

Person: Well, I started smoking 15 years ago. I tried several times to quit smoking and even managed to go two months without smoking. I have heard about consultations out there, but I want none of that. It is about my willpower...to stop smoking.
Professional: You should really seek professional support to get a better chance to quit smoking.

The following two dialogues illustrate the previous concepts based on the communication design proposed by Rollnick et al. (2005):

Dialogue 1

Professional: *Your test result shows that glucose levels in your blood are raised today. This means that you really need to watch your diet. Have you thought about adjusting this?*

Person: *Well, it is not that easy. I have tried, but you know what it's like. I mean, it's not that easy with my job, driving around in a rush, and you know, you just have to grab some food at lunch and keep going.*

Professional: *Could you bring your own lunch with you…?*

Person: *I could do that, but it's so busy in the morning, just getting us all out of the house, and then I stop in a cafe anyway at lunch, so I would then have to avoid the easy option of just getting a roll and feeling full and ready for action.*

Professional: *Well, you are treating this as your top priority, right?*

In Dialogue 1, the professional uses an informing mode drawing on a rigid interaction routine, making assumptions about the diet, uses a predominance of closed questions and resorts to leading or directive questions. This may elicit resistance to change and guilt and is unlikely to generate opportunities for changing behaviour.

Dialogue 2

Professional: *Your test result shows that glucose levels in your blood are raised today. I wonder what sense you make of this?*

Person: *I don't know. It's hard to live 24/7 with diabetes, I'm so busy, and it's another thing to worry, the blood sugar levels.*

Professional: *I completely understand. Everyday life can't stop because you have diabetes [empathic response].*

Person: *Yes, exactly, but I know I need to be careful.*

Professional: *In what way?*

Person: *I need to watch my diet and get more exercise. I know that, but it's not that easy.*

Professional: *What might be manageable for you right now?*

Person: *It's got to be exercise, but please don't expect great things from me.*

Professional: *Well, a change in diet or exercise may be of help. How might you succeed with exercising more?*

Dialogue 2 corresponds to a **guiding communication style** or a **patient-centred dialogue**. Asking open questions elicits the person's perspective on self-management behaviours. Listening is used to convey an understanding of the person's experience and to encourage further exploration. Informing is combined with asking, to encourage choice and promote autonomy. This option is more likely to generate opportunities for changing behaviour.

Other crucial interpersonal communication skills involve using positive non-verbal and body language (e.g. visual contact, framing shoulders with the other, leaning the torso slightly forward), optimising verbal language (as described in the next section) and managing the exchange to provide opportunities for the person to speak.

5.2.2 Optimising Verbal Language

Person-centred care has implications for the language used by professionals when communicating with persons with chronic diseases. For example, the biomedical lexicon can hinder comprehension for some persons. As explained in the section on "health literacy", using plain language minimises the risk of misunderstanding. Beyond this aspect, recommendations on the use of language to communicate with persons with diabetes and obesity have been issued (Banasiak et al., 2020; Cooper et al., 2018; Dickinson et al., 2017; Speight et al., 2012; Speight et al., 2021), to encourage terms promoting positive interactions and, subsequently, positive outcomes.

Table 5.2 provides examples of preferred terms for behaviour change support interactions in chronic disease based on this work. In their scoping review, Lloyd et al. (2018) elegantly articulated arguments in favour and against the use of "person-first language" ("person with diabetes") instead of a disease-first language ("diabetic"). While some persons may find it acceptable to refer to themselves as "diabetics", others find it offensive or harmful; professionals have the responsibility to use language that respects everyone's preferences. As Speight et al. (2021) pointed out, people are rarely offended by being referred to as a "person".

5.2.3 Coping Skills to Manage Chronic Disease

When dealing with a chronic disease, it is common to experience negative emotions related to the disease onset, progress and treatments, as well as psychosocial implications, e.g. social isolation and difficulties in engaging in daily activities. This is illustrated by the concept of diabetes distress, introduced in 1995 to designate "the negative emotional or affective experience resulting from the challenge of living with the demands of diabetes" (Skinner et al., 2020). There is compelling evidence that diabetes distress is common, affecting roughly one in three persons living with type 2 diabetes, and is a barrier to emotional well-being, self-care and diabetes management. Consequently, monitoring diabetes distress is advocated by many clinical guidelines (Skinner et al., 2020).

It is also common for persons with chronic disease to experience various lapses when trying to maintain self-management behaviours, such as physical activity, a healthy diet or following prescribed treatment. To deal with these stressful situations, people can implement various coping strategies.

Table 5.2 Optimising the use of words in person-centred behaviour change support

Avoid	Prefer	Rationale
Diabetic/asthmatic/ hypertensive/etc.; victim of…; sufferer; patient	Person living with diabetes, asthma, hypertension, etc.; person with diabetes, asthma, hypertension, etc.; diagnosed with diabetes, asthma, hypertension, etc.	The labels "diabetic/asthmatic/ hypertensive", etc. may be offensive for some persons, as it defines them based on a condition. "Victim of…" and "sufferers" positions people as passive and helpless rather than empowering them to live with the disease.
Normal, healthy	Person living without diabetes, asthma, hypertension, etc.	Referring to people without chronic disease as "normal" implies that those living with chronic disease are abnormal, which is stigmatising.
In denial	Finding it difficult	Persons living with chronic diseases may take time to adjust to this reality; not everyone adjusts at the same pace. Labelling them as "in denial" is judgmental and unlikely to lead to a co-operative relationship with the professional.
Unmotivated, unwilling	Concerned with Has other priorities right now	Obstacles to chronic disease self-management may sometimes be perceived as insurmountable or not worth the effort. This should be respected and support offered. Labelling is judgmental and unlikely to lead to a co-operative relationship with the professional.
Adherent/ non-adherent Compliant/ non-compliant	Terms focusing on what the person can do, e.g. "takes medication as agreed most of the time" and "eats fruits and vegetables some days per week"	Language should not imply that the person follows orders or passively does what is told by professionals. Supporting self-management entails collaboration between the professional and the person, considering personal preferences and priorities. It is more helpful to explore barriers to self-management than labelling the person.
Difficult/ challenging patient	Difficult/challenging situation	Chronic disease management can be demanding; terms should describe the situation and not the person. Labelling is judgmental and unlikely to lead to a co-operative relationship with the professional.

Based on Banasiak et al. (2020), Cooper et al. (2018), Dickinson et al. (2017), Speight et al. (2012) and Speight et al. (2021)

The American Psychological Association defines coping as "the use of cognitive and behavioural strategies to manage the demands of a situation when these are appraised as taxing or exceeding one's resources or to reduce the negative emotions and conflict caused by stress" (from https://dictionary.apa.org/coping).

The individual can adopt various coping strategies to self-manage chronic disease resulting from the cognitive and emotional representations associated with the stressors. These coping strategies can be positive (i.e. adaptive), e.g. taking time to exercise in the middle of a hectic day, or negative (i.e. maladaptive, avoidant), e.g. not asking for support when decisions on medication-taking are not being met, binge drinking or overeating.

One of the main distinctions of coping strategies is between problem-focused coping and emotion-focused coping (Lazarus & Folkman, 1984). As the name suggests, problem-focused coping aims to remove or reduce the cause of stress by the person through, e.g. problem-solving techniques (such as coping or barrier planning, a common behaviour change technique previously described), better time management or support from others. Emotion-focused coping includes those strategies used to regulate the person's negative emotions, such as fear, anxiety, sadness or frustration; examples are distraction, mindfulness and relaxation. Please note that distraction is a coping strategy that fits various categorisations of coping strategies. Figure 5.3 presents more examples of problem-focused and emotion-focused coping strategies.

Another common distinction is between approach and avoidance coping. Approach coping is any behaviour, cognitive or emotional activity that directly deals with the stressor or threat, such as problem-solving using if-then plans ("If I feel too tired, I will do a shorter exercise session and with a less intensive pace"). Avoidance coping refers to any behaviour, cognitive or emotional activity taken to

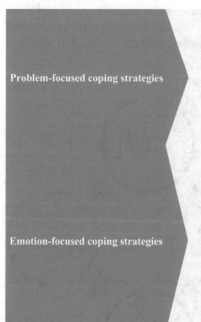

Problem-focused coping strategies

Creating a to-do list (e.g., steps needed to discuss treatment options in the next consultation with the prescriber)
Establishing healthy boundaries (e.g., not committing to unreasonable goals in physical activity or diet to please a health professional)
Avoiding stressful situations (e.g., choosing not to spend time with relatives that pass on judgements on how the person deals with chronic disease; changing to a more empathic health professional)

Emotion-focused coping strategies

Distraction (e.g., engaging in a hobby; exercise; focus on a task, such as cleaning the house, cooking a meal, or reading a book)
Relaxation (e.g., playing with a pet; practice breathing exercises; squeezing a stress ball; using a relaxation app; enjoying things that make the person feels good, such as doing the hair and taking a bath, drinking tea; positive self-talk)
Mindfulness (e.g., listing things for which the person feels grateful; meditation; picturing a "happy place"; looking at pictures of people and places that bring joy)
Journaling (e.g., keep a diary that explores thoughts and feelings surrounding self-managing chronic disease, counteracting negative feelings with potential solutions, things the person appreciates in life, or things that give hope to the person)

Fig. 5.3 Practical examples of problem- and emotion-focused coping strategies

avoid the threat. It can be maladaptive, such as denial of diagnosis or progress of a disease or withdrawal from a beneficial treatment; in other instances, it can be helpful, for example, distraction (doing something else to avoid thinking about the problem in a given situation can help in reducing high levels of distress).

When managing chronic disease, individuals should be flexible and use different types of coping skills according to the characteristics of the situation, such as the level of control the individual has over the situation or if it leads to powerful emotional reactions.

Professionals can support persons in understanding the type of coping skills they tend to use for different stressful situations and deal with negative emotions, if these are adaptive and maladaptive, and facilitate the acquisition and enactment of coping skills described in Fig. 5.3. Each person may need to experiment with various coping strategies to discover which ones work best as ongoing events and life circumstances change, and come up with his or her own toolkit of strategies.

When exploring and training individuals in coping skills, professionals must use the communication skills addressed in this chapter, namely, non-judgemental and empathic communication.

5.2.4   Structuring the Interaction: The ABCD Approach

Smith et al. (2000) have proposed evidence-based guidelines for patient-centred communication. Their approach has been adapted in this book for providing behaviour change support. In the interest of simplicity, our adaptation has been coined ABCD, as it encompasses four sequential stages, detailed in Fig. 5.4, which in turn contain several steps.

Fig. 5.4 The ABCD approach for structuring each behaviour change interaction

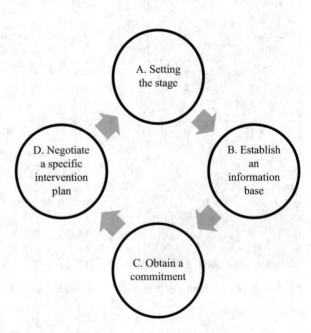

Table 5.3 Purpose of the ABCD stages: first encounter

Stage	Purpose
A. Setting the stage.	Ascertain the receptiveness to discuss behaviour change
B. Establish an information base.	Discuss the behaviours to be changed and factors influencing them
C. Obtain a commitment.	Engage in the decision of changing behaviour informed by realistic expectations
D. Negotiate an intervention plan.	Collaboratively setting up an intervention plan

Table 5.4 "Setting the stage" steps

Long interventions	Brief interventions
1. Welcome the person	1. Welcome the person
2. Introduce self and identify the specific role	7. Agree on the agenda
3. Ensure privacy	
4. Remove barriers to communication	
5. Ensure comfort	
6. Put the person at ease	
7. Agree on the agenda	

These sequential stages apply to each behaviour change interaction, either in brief or long-term interventions, although not all steps subsumed in the different stages are necessarily applicable. Brief interventions are usually opportunistic and shorter, and therefore within each stage, some steps may be omitted, as explained in the following sections.

Long-term interventions involve follow-up, which may also be part of brief interventions. Interactions after the first encounter resort to the same sequential stages, as depicted by the circular nature of this approach (Fig. 5.4).

The ABCD approach is meant to aid professionals in flexibly structuring their behaviour change interactions, considering both the context and the person, and is not intended to be used prescriptively. Table 5.3 outlines the purpose of each stage.

5.2.4.1 First Encounter

A. Setting the stage

Setting the stage is probably the most challenging component of effective behaviour change interventions. Fisher et al. (2017) highlighted that professionals should be prepared for a different interactive mindset and to make an effort to slow the conversation pace. Additionally, professionals need to re-orient themselves to motivational needs and obstacles, i.e. moving from an information delivery focus to listening and reflecting on conversation. It is not about extra time but addressing the nuances of the person's engagement by supporting and encouraging their motivation to change (Fisher et al., 2017).

Setting the stage in the present model comprises seven steps, detailed in Table 5.4. It may be unfeasible to implement these seven steps in brief interventions due to their shorter and opportunistic nature, and therefore, two steps (welcoming the person and agreeing on the agenda) are suggested as a priority.

When welcoming the person, the professional should ask how the person wishes to be addressed, by first or last name, sometimes including titles. It is generally a good practice to use the surname to start with, although this may vary between countries and cultures.

Steps 2 to 6 intend to develop rapport, commonly defined as the level of connection between persons communicating bidirectionally (Box 5.10). Furthermore, a relationship based on mutual trust is expected when rapport is established. Developing rapport with the person is essential since it builds meaningful conversations and promotes the willingness to embrace different points of view.

Box 5.10 Example of a Dialogue Initiation and Rapport Building

Professional: Good morning. You must be Paul.

Person: That's right.

Professional: Come on in, and make yourself comfortable, Paul. Is it all right to call you Paul?

Person: I would prefer Mr. Jonhson.

Professional: Sure. Let me close the door, Mr. Johnson, so we can have a quiet talk. My name is Jeff, and I provide counselling on health-related behaviours. Your wife mentioned that you were thinking of quitting smoking.

Person: Well, my wife suggested this appointment…

Professional: I'm glad you had the time to visit us today, Mr. Johnson. Your wife expressed concern about your increasing breathing difficulties.

Person: She cares about me a lot.

Professional: That's nice to know. During your time with us today, I would like you to feel at ease. If, for any reason, you want to end the conversation, that's perfectly fine. We can resume it on another occasion if you prefer.

Person: I understand.

Professional: So, our conversation will be about ways to help you give up smoking. In general, this does not take more than 20 minutes. Would that be OK for you?

Person: Perfectly fine. Thank you for your time.

The last point, agreeing on the agenda, in this case comprising health behaviour change, should happen when both the professional and the person are bonded in the conversation, even for other reasons. In brief interventions, it may be feasible to use a warm welcome or quick social talk to facilitate this bond. In essence, agreeing on the agenda means deciding whether the person is receptive to discussing behaviour change. Directive conversation to persuade the person to change behaviour based on the professional's agenda, disregarding the person's priorities, preferences, beliefs and resources, is not congruent with a person-centred approach.

Four techniques described in the literature to initiate a health behaviour change talk (Albury et al., 2020) are helpful to agree on the agenda and decide whether the interaction can proceed to the next stage (B. Establish an information base). The first three techniques described below should be preferred.

Person-initiated discussion is a technique where the person initiates the talk by asking directly for advice or raising a concern (Box 5.11). The professional should capitalise on this situation, as it facilitates receptiveness.

> **Box 5.11 Example of Person-Initiated Discussion**
>
> Person: You know, smoking is a health issue for me. I have started to have shortness of breath when I exercise.
> Professional: OK. Have you considered quitting smoking?

Direct questioning by professionals (e.g. do you smoke?) may lead to acknowledging the risk behaviour and eliciting helpful information, such as disclosing attempts to change behaviours or providing explanations for not doing so. Professionals can resort to the information elicited to inform the subsequent discussion.

Non-personalised initiation is another technique that is less likely to generate resistance. As described in Box 5.12, this approach consists of establishing a generic statement on the behavioural problem, securing agreement on this statement from the person and then moving to a more personalised discussion.

> **Box 5.12 Example of Non-personalised Initiation**
>
> Professional: Smoking is a public as well as an individual health concern. Implications of smoking are well established, such as damage to the lungs and the heart.
> Person: You know, smoking is a health issue for me. I have started to have shortness of breath when I exercise.
> Professional: OK. Have you considered quitting smoking?

As illustrated in Box 5.13, linking health behaviours to a medically relevant concern is expected to facilitate the interaction by emphasising its salience for the person. However, this technique has varying effectiveness; it has been shown to elicit resistance even when the link between the health concern and health behaviour had relevance to the person. Resistance has been defined as a response that halts the conversation progressivity, ranging from no response, a minimal response, or not displaying alignment with behaviour change (Albury et al., 2020).

Box 5.13 Example of Linking Health Behaviour to a Medically Relevant Concern

Person: You know, I cannot get pregnant right now. So, I have to take my contraceptive pills.
Professional: Do you still smoke?
Person: Yeah.
Professional: Following regular hormonal contraception is generally accepted as an effective and safe option. At the same time, smoking and taking birth control pills brings additional health risks.

Caution should be exercised when associating a person's disease with health behaviours, as several other determinants play a role in the health state. These determinants include healthcare, genetic vulnerability, socio-economic characteristics and environmental and physical influences, mainly beyond the person's control (Naughton, 2018). Linking a person's disease with individual health behaviours implies being blamed for the disease, which can lead to resistance to change and should therefore be avoided.

Initiating a change in topic is recommended to manage resistance during behaviour change talks (Albury et al., 2020); it can be equally helpful to deal with displays of resistance when "Setting the stage", as exemplified in Box 5.14.

Box 5.14 Example of Change in the Topic When Setting the Stage

Professional: Do you still smoke?
Person: Yeah.
Professional: There are some strategies that really help with quitting…
Person: [5 sec silence]
Professional: Well… so… how're you getting along with your pedometer? Seems okay?
Person: No problem… I'm enjoying tracking of my progress
Professional: Tracking your progress… good.

After confirming the receptiveness to discuss behaviour change, the professional can suggest continuing the conversation to collect more information, which should be timewise, i.e. based on the professional and the person's availability, as illustrated in Box 5.15.

Box 5.15 Example of a Dialogue on Time Limits When Setting the Stage

Professional: So, I see that you have been thinking about quitting smoking. Should we discuss this?
Person: Well, I only have about 5 minutes, actually…
Professional: That's fine, I am also pressured today. What about starting to discuss, and you can return on another day?
Person: Sounds good!

Table 5.5 "Establish the information base" steps

Long-term and brief interventions
1. Identify self-management needs in relation to target behaviours
2. Elicit behaviour determinants (e.g. COM-B model – See Sect. 5.2.3 for details)

B. Establish an information base

Establishing the information base comprises two steps, detailed in Table 5.5. Although the depth of information collected may vary, these two steps apply to long-term and brief interventions. For instance, discussing factors that influence a behaviour is required in a brief intervention to inform the plan, but it will be less comprehensive than in a long-term intervention due to time constraints.

Put simply, identifying self-management needs answers the question "what target behaviour(s) need to be changed?" and leads to prioritisation if more than one risk behaviour emerges. As explained in Sect. 3.2, assessing the person's behaviour relies, often, on an interview. The communication should begin with open-ended questions, using non-focusing open-ended skills: silence, neutral utterances and non-verbal encouragement. These can be followed by focusing on open-ended inquiry, if needed, to get the person talking (e.g. parroting and summarising). Further, closed-ended questions can be used for clarification and additional information can be obtained from sources such as non-verbal cues, physical characteristics, autonomic changes and the environment.

Box 5.16 Example of Dialogue to Establish an Information Base

Professional: I see that you have asthma. What other issues would you like to discuss, in addition to smoking?

Person: Well… I don't use my inhalers exactly as prescribed, but I'm doing fine, so I don't regard that as an issue.

Professional: Hum, hum.

Person: Really, smoking is a health issue for me. I started to have shortness of breath when I exercise.

Professional: Right. Do you regard quitting smoking as a priority?

Person: Not sure if it is a priority, I'm not doing much.

Professional: I'm sure you did what you could. How many cigarettes do you smoke per day?

One note on non-verbal information. While body presentation and language are well recognised and possible to interpret correctly based on the social and cultural background, autonomic changes such as cold sweating and abdominal discomfort are not easily detected.

The second step, eliciting behaviour determinants, should follow the same communicational approach when an interview is used: starting with open-ended questions on what stops the person from adopting the target behaviour and what facilitates its adoption. It may be helpful to focus questions on barriers and facilitators in adopting a target behaviour (e.g. quitting smoking, being physically active) instead of exploring factors influencing the person's current behaviour (e.g. what makes the person smoke, what makes the person sedentary). For assessing behaviour determinants using measures refer to Sect. 4.1.2.

The focus on the person's relevant factors improves rapport and trust. Closed questioning is also used to confirm the information elicited. Eliciting behaviour determinants in long and brief interventions through interviews is exemplified below; in the former, there is time for more exploration through listening skills, such as parroting and summaries. To save time, paraphrasing may be preferable in brief interventions instead of parroting, which encourages the person to continue talking.

Box 5.17 Example of Eliciting Behaviour Determinants in a Brief Intervention

Professional: Can you think of issues that prevent you from stopping smoking?
Person: Well… stress, really.
Professional: So, I understand that stress gets in the way of your will to stop smoking.

Box 5.18 Example of Eliciting Behaviour Determinants in a Long-term Intervention

Professional: Can you think of issues that prevent you from stopping smoking?
Person: Well… stress, really.
Professional: Stress….
Person: I deal with a lot of pressure at work, and it's a way of unwinding. And it piles up, I get home, the kids and all that, you want to be in your best for them, so after dinner it also helps me to unwind.
Professional: Right. Are there other factors that get in the way of your will to stop smoking?
Person: Now that you mention that… I never had support. I have heard about consultations out there, but I don't know how to access them.
Professional: Correct me if I'm wrong… the main reasons for not stopping smoking are stress and not knowing how to access smoking cessation consultations.

Assessing the person's readiness to change involves ascertaining the degree of motivation which influences the behaviour, as illustrated by the COM-B model. Establishing motivation is important in the first and follow-up interactions, mainly if behaviour change is not being unattained.

The motivation ruler is a 1–10 scaling exercise to help the person think about the target behaviour and articulate reasons for changing. As the person verbalises these reasons, they may become more natural. To maintain a cooperative relationship with the person, questions such as "Why is it not higher?" or "Why x?" should be avoided. This exchange may be regarded as confrontational and raise resistance. It is vital to promote an empathic climate through positive communication. In the end, the professional should help the person summarise the reasons underlying the performance of a behaviour. An examples is presented in Box 5.19 regarding smoking cessation, focusing on communication aspects. [1]

Box 5.19 Example of a Dialogue Using the Motivation Ruler

Professional: On a scale of 1–10, how important for you is stopping smoking in the next three months?
Person: 5.
Professional: OK. What was important to you when you decided on that score?
Person: You know, smoking is a health issue for me. I started to have shortness of breath when I exercise.
Professional: I see. Have you considered other issues regarding smoking?

C. Obtain a commitment

Obtaining a commitment comprises two steps, detailed in Table 5.6. Implementing these two steps in brief interventions may not be feasible due to their shorter and opportunistic nature. Step 1 – discussing expectations – is not vital to define the intervention plan but may be helpful to smooth implementation. Box 5.20 illustrates a dialogue to reaffirm commitment in a brief intervention.

Table 5.6 "Obtain a commitment" steps

Long interventions	Brief interventions
1. Discuss expectations for success 2. Reaffirm commitment	2. Reaffirm commitment

---

Box 5.20 Example of Obtain a Commitment in a Brief Intervention

Professional: Now that you told me about what prevents you from quitting smoking… we can discuss your options and adapt them to your needs and preferences. How does this sound?
Person: That's very much appreciated.

---

[1] Additional details can be found in textbooks, such as the following reference: Miller, W.R., & Rollnick, S., (2012). Motivational Interviewing: Helping People Change (Applications of Motivational Interviewing). London: Guildford Press.

Ascertaining the expectations for success entails understanding the person's views on the process (e.g. what it is about and how to achieve it), as exemplified in Box 5.21, and providing a different perspective if expectations appear unrealistic. Before reaffirming commitment, the professional may need to address emotion(s) empathically. Filling the person's information gaps may also be required to reaffirm commitment to decisions and actions in behaviour change. In long interventions, the professional has room to provide more detailed explanations on the process of behaviour change, if pertinent.

Box 5.21 Example of Obtain a Commitment in a Long-term Intervention

Professional: Let's discuss your expectations about quitting smoking. What do you think is ahead of you?

Person: Well, I tried several times to quit smoking and even managed to go two months without any cigarettes. Then I went back again. I don't think it's easy...

Professional: As you said, many people find it challenging. We are here to offer support throughout the process! We can discuss your options and adapt them to your needs and preferences. How does this sound?

Person: That's very much appreciated.

Table 5.7 "Negotiate an intervention plan" steps

Long interventions	Brief interventions
1. Discuss BCTs addressing behaviour determinants and decide collaboratively on the intervention plan	1. Discuss BCTs addressing behaviour determinants and decide collaboratively on the intervention plan
2. Closing	2. Closing

D. Negotiate an intervention plan

The steps subsumed in the "Negotiating an intervention plan" stage are detailed in Table 5.7. Due to their opportunistic nature, it may be unfeasible to implement follow-up in brief interventions.

The professional should keep in mind the communication skills and strategies already described (e.g. questioning, active listening, change in topic to manage resistance). Linking health behaviours to a salient concern may be helpful at at this stage to address resistance. Asking for feedback is a relevant way to check the person's rapport and pay attention to non-verbal communication signs.

Negotiating a specific plan can draw on shared decision-making models' strategies, such as the three-talk model (Elwyn & Durand 2017), which includes "team talk", "option talk" and "decision talk".

"Team talk" refers to making sure the person knows that reasonable options are available, to provide support when making the person aware of choices and to elicit the person's goals to guide decision-making processes.

"Option talk" refers to providing more detailed information about options and comparing alternatives using risk communication principles. While this has been established for comparing alternatives in drug therapy, it is less settled on what concerns different possible strategies to change behaviour.

"Decision talk" refers to arriving at decisions that reflect the informed preferences of persons, guided by the experience and expertise of health professionals.

Box 5.22 offers an example of applying the three-talk model to behaviour change in medication-taking; this example illustrates honesty about what is known and explores the person's understanding, reactions and opinions about the information.

Box 5.22 Example of a Three-Talk Dialogue in Medication-Taking

Person: As I said, it's not that I don't want to take my blood pressure pill; it's just that I have too much going on some days….

Professional: So, let's work as a team to make a decision that suits you best. One option that appears suitable is setting up an alarm when you choose to take your blood pressure pill, for example, using your mobile phone. You said you take the pill in the morning?

Person: Yeah, that's right.

Professional: OK. Other options are having your pillbox on the table where you have breakfast to see them or bringing a blister pack around with you in your purse/wallet. Are there any children in your house?

Person (laughs): Not really. They got off to university.

Professional: Oh, that's lovely. Having medicines within reach of children would not be a good idea. So, what do you think of these three options?

Person: They sound good, but whether they work for me is a different story. I already have my pillbox on the kitchen counter.

Professional: So, you're telling me that you forget to take your blood pressure pill in the morning, but having the medicine at sight doesn't work. We don't really know whether what works best in terms of comparing these options. It depends on each person's preference and resources.

Person: Right….

Professional: If you are the type of person who uses a mobile phone, you might want to set up an alarm.

Person: Well… I tend to stop alarms and then still forget.

Professional: If the alarm doesn't work for you currently, then maybe add a post-it to the fridge door?

Person: I think I could give it a try; after a couple of days, I might not notice the post-it anymore….

Professional: OK, let's go with your choice, then! When would you like to start sticking the post-it on the fridge's door?

Person: I'll give it a try this week.

Professional: Great! You will tell me whether it works in our next encounter.

Active listening is a central component of deliberation, i.e. the process where the person becomes aware of the choice, considers the pros and cons of the options and assesses their practical and emotional implications.

Deliberation begins as soon as awareness about options develops. The process is iterative, as options have been described and understood. Deliberation encompasses the need to work collaboratively with professionals and may involve the person's broader networks.

The person may need time and support to reflect on preferences and practicalities. Therefore, several encounters with the professional may be required, not necessarily face to face, and may include decision aids and discussions with others.

In addition to the three-talk model, which applies to any preference-sensitive decision in healthcare, the literature offers specific recommendations to guide communication during health behaviour change interventions (Albury et al., 2020). Collaborative talk, exemplified in Box 5.22, is associated with uptake displays. It requires inviting and accommodating the person's perspective and presenting decisions as their choice. This strategy may facilitate engagement when a display of resistance emerges, and the professional chooses to continue the conversation. As already explained, initiating a change in the topic is a technique that also helps managing displays of resistance at this stage, as evidenced in Box 5.23.

Box 5.23 Example of a Change in the Topic When Negotiating an Intervention Plan

Person: As I said, it's not that I don't want to take my blood pressure pill; it's just that I have too much going on some days....

Professional: Bringing a blister pack around with you in your purse may help.

Person: Everybody keeps telling me that but it doesn't work for me.

Professional: OK. So, let's work as a team to make a decision that suits you best. One option that appears suitable is setting up an alarm when you choose to take your blood pressure pill, for example, using your mobile phone. You said you take the pill in the morning?

Person: Yeah, that's right.

Deciding collaboratively on the intervention includes agreeing on BCTs, to be implemented during encounters or techniques that the person can self-enact. Finalise this step includes confirming the person's understanding and reaffirming the plan Box 5.23.

The encounter closure should include setting dates/times for a follow-up visit, usually in the case of long interventions. Brief interventions do not necessarily encompass follow-up due to their opportunistic nature. Avoiding repetition of agreed actions contributes to expediting closing.

Table 5.8   Purpose of the ABCD stages: in follow-up encounters

Stage	Purpose
A. Setting the stage	Ascertain the receptiveness to discuss the intervention plan
B. Establish an information base	Discuss the implementation of the intervention plan
C. Obtain a commitment	Gauge the person's engagement in behaviour change
D. Negotiate an intervention plan	Collaboratively reviewing the intervention plan

5.2.4.2   Follow-Up Encounters

Interactions after the first encounter resort to the same ABCD stages; Table 5.8 details the purpose of each stage in follow-up encounters.

It is essential to monitor the intervention plan in these encounters, i.e. gather information to assess progress and adjust the plan accordingly.

The encounter should start by setting the stage. In follow-up encounters, agreeing on the agenda means deciding whether the person is receptive to discuss the intervention plan. Displays of resistance can be managed by initiating a change in the topic or offering the opportunity of a subsequent encounter.

In stage B, the professional should gather information regarding the agreed plan, using the communication strategies already discussed (e.g. open to close questioning, active listening). Discussing the intervention plan's implementation involves ascertaining how it is working regarding the application of behaviour change techniques and the outcomes achieved. If the plan is not working, either because the person is struggling with self-enactable BCTs or expected outcomes are not achieved, it is helpful to review BCTs application and behaviour determinants, as uncovered or additional barriers may emerge. An empathic and legitimising conversation, as described in Box 5.24, focusing on achievements, current or future, is more empowering than emphasising failures.

Box 5.24 Example of an Empathetic Dialogue When Establishing an Information Base (Follow-up Encounter)

Person: Well, I was unable to cut down the number of daily cigarettes. The last two weeks were very stressful for me.

Professional: Good that you are still motivated to quit smoking. I completely understandable that it is hard to reduce the number of cigarettes under stress. Would you like to talk more about this?

If the plan is working, the professional should positively reinforce the person's efforts when reviewing their actions to change behaviour. If only small successes were achieved, the talk should also focus on reviewing BCTs application and behaviour determinants without blaming the person.

Box 5.25 Example of a Non-blaming Dialogue When Establishing an Information Base (Follow-up Encounter)

Person: Well, I'm smoking just once a day when I'm feeling very stressed.
Professional: Excellent! Many congrats on achieving this outcome. Should we discuss strategies to reduce stress or alternative ways of avoiding that one cigarette?

Maintenance should be discussed when a target behaviour is achieved, self-management needs identified again and the process restarted for another target behaviour, if pertinent. Guidance to address multi-behaviour change interventions is still in a nascent phase.

Stage C, "Obtain a commitment", involves discussing the person's expectations and reaffirming the intention of engaging in behaviour change.

In Stage D the plan should be reaffirmed or reviewed collaboratively through the three-talk model or techniques such as inviting and accommodating the person's perspective and presenting decisions as the person's choice, already described. Reviewing the plan should consider information gathered in stage B and his or her engagement in behaviour change.

Key Points

- At a meso- and micro-level patient empowerment encompasses approaches such as patient-centredness and shared-decision making, which contribute to increase the capacity of persons living with chronic disease to think critically and make independent and informed decisions.
- Key aspects of person-centredness care are considering individual preferences, needs and values, being respectful of and responsive to those preferences, needs and values, and thus fully engaging the person in the intervention.
- Shared decision-making is a collaborative process in which a (healthcare) professional works together with a person to make health-related decisions based on evidence and individual choices.
- Education aims to increase knowledge and ensure that the person makes excellent and independent use of that knowledge. The "Elicit-Provide-Elicit" technique promotes personalised education.
- Health literacy involves not only comprehending health information, but also the ability to access it and use it effectively.
- To effectively support behaviour change professionals need relational and communication skills, including the use of open-ended questions, active listening and empathetic responding. Talking to persons with chronic disease should preferably make use of language that is non-judgmental, blame-free and that empowers.
- Professionals can assist persons with chronic disease to develop Coping skills (e.g. problem-focused coping and emotion-focused coping) to deal with negative emotions and psychosocial implications associated with their conditions.
- The ABCD approach is a framework to structure behaviour change interactions, it facilitates engagement through person-centred communication and shared-decision making.

References

Agbadjé, T. T., Elidor, H., Perin, M. S., Adekpedjou, R., & Légaré, F. (2020). Towards a taxonomy of behavior change techniques for promoting shared decision making. *Implement. Sci., 15*(1), 67. https://doi.org/10.1186/s13012-020-01015-w

Albury, C., Strain, W. D., Le Brocq, S., Logue, J., Lloyd, C., & Tahrani, A. (2020). The importance of language in engagement between healthcare professionals and people living with obesity: a joint consensus statement. *Lancet Diabetes Endocrinol., 8*(5), 447–455. https://doi.org/10.1016/S2213-8587(20)30102-9

Anderson, R. M., & Funnell, M. M. (2010). Patient empowerment: myths and misconceptions. *Patient Educ. Couns., 79*(3), 277–282. https://doi.org/10.1016/j.pec.2009.07.025

Banasiak, K., Cleary, D., Bajurny, V., Barbieri, P., Nagpal, S., Sorensen, M., et al. (2020). Language matters–a diabetes Canada consensus statement. *Can. J. Diabetes, 44*(5), 370–373. https://doi.org/10.1016/j.jcjd.2020.05.008

Beauchamp, T., & Childress, J. (2019). Principles of biomedical ethics: marking its fortieth anniversary. *Am. J. Bioeth., 19*(11), 9–12. https://doi.org/10.1080/15265161.2019.1665402

Bull, E. R., & Dale, H. (2021). Improving community health and social care practitioners' confidence, perceived competence and intention to use behaviour change techniques in health behaviour change conversations. *Health Soc Care Community, 29*(1), 270–283. https://doi.org/10.1111/hsc.13090

Cooper, A., Kanumilli, N., Hill, J., Holt, R. I. G., Howarth, D., Lloyd, C. E., Kar, P., Nagi, D., Naik, S., Nash, J., Nelson, H., Owen, K., Swindell, B., Walker, R., Whicher, C., & Wilmot, E. (2018). Language matters. Addressing the use of language in the care of people with diabetes: position statement of the English Advisory Group. *Diabet. Med., 35*(12), 1630–1634. https://doi.org/10.1111/dme.13705

Cupples, M. E., Cole, J. A., Hart, N. D., Heron, N., McKinley, M. C., & Tully, M. A. (2018). Shared decision-making (SHARE-D) for healthy behaviour change: a feasibility study in general practice. *BJGP Open, 2*(2). https://doi.org/10.3399/bjgpopen18X101517

Dickinson, J. K., Guzman, S. J., Maryniuk, M. D., O'Brian, C. A., Kadohiro, J. K., Jackson, R. A., et al. (2017). The use of language in diabetes care and education. *Diabetes Care, 40*(12), 1790–1799. https://doi.org/10.2337/dci17-0041

Elwyn, G., & Durand, M. A. (2017). *Mastering shared decision making: the when, why and how*. EBSCO Health. (https://health.ebsco.com)

Elwyn, G., Frosch, D., Thomson, R., Joseph-Williams, N., Lloyd, A., Kinnersley, P., et al. (2012). Shared decision making: A model for clinical practice. *J. Gen. Intern. Med., 27*(10), 1361–1367. https://doi.org/10.1007/s11606-012-2077-6

Fisher, L., Polonsky, W. H., Hessler, D., & Potter, M. B. (2017). A practical framework for encouraging and supporting positive behaviour change in diabetes. *Diabet. Med., 34*(12), 1658–1666. https://doi.org/10.1111/dme.13414

Funnell, M. M., Anderson, R. M., Arnold, M. S., Barr, P. A., Donnelly, M., Johnson, P. D., et al. (1991). Empowerment: an idea whose time has come in diabetes education. *Diabetes Edu, 17*(1), 37–41. https://doi.org/10.1177/014572179101700108

Gültzow, T., Zijlstra, D. N., Bolman, C., de Vries, H., Dirksen, C. D., Muris, J. W. M., Smit, E. S., & Hoving, C. (2021). Decision aids to facilitate decision making around behavior change in the field of health promotion: a scoping review. *Patient Educ. Couns., 104*(6), 1266–1285. https://doi.org/10.1016/j.pec.2021.01.015

Hamelinck, V. C., Bastiaannet, E., Pieterse, A. H., Jannink, I., van de Velde, C. J. H., Liefers, G. J., & Stiggelbout, A. M. (2014). Patients' preferences for surgical and adjuvant systemic treatment in early breast cancer: a systematic review. *Cancer Treat. Rev., 40*(8), 1005–1018. https://doi.org/10.1016/j.ctrv.2014.06.007

Holmström, I., & Röing, M. (2010). The relation between patient-centeredness and patient empowerment: a discussion on concepts. *Patient Educ. Couns., 79*(2), 167–172. https://doi.org/10.1016/j.pec.2009.08.008

Kayser, L., Karnoe, A., Duminski, E., Somekh, D., & Vera-Muñoz, C. (2019). A new understanding of health related empowerment in the context of an active and healthy ageing. *BMC Health Serv. Res., 19*(1), 1–13. https://doi.org/10.1186/s12913-019-4082-5

Lau, D. H. (2002). Patient empowerment--a patient-centred approach to improve care. *Hong Kong Med. J., 5*, 372–374. (PMID: 12376717).

Lazarus, R. S., & Folkman, S. (1984). *Stress, appraisal and coping.* Springer.

Mead, N., & Bower, P. (2002). Patient-centred consultations and outcomes in primary care: a review of the literature. *Patient Educ. Couns., 48*(1), 51–61. https://doi.org/10.1016/S0738-3991(02)00099-X

Michie, S., van Stralen, M. M., & West, R. (2011). The behaviour change wheel: a new method for characterising and designing behaviour change interventions. *Implement. Sci., 6*(1), 42. https://doi.org/10.1186/1748-5908-6-42

Mokhar, A., Topp, J., Härter, M., Schulz, H., Kuhn, S., Verthein, U., & Dirmaier, J. (2018). Patient-centered care interventions to reduce the inappropriate prescription and use of benzodiazepines and z-drugs: A systematic review. *PeerJ.* 6:e5535. https://doi.org/10.7717/peerj.5535

Rollnick, S., Butler, C. C., McCambridge, J., Kinnersley, P., Elwyn, G., & Resnicow, K. (2005). Consultations about changing behaviour. *BMJ, 331*(7522), 961–963. https://doi.org/10.1136/bmj.331.7522.961

Scholl, I., Zill, J. M., Härter, M., & Dirmaier, J. (2014). An integrative model of patient-centeredness-A systematic review and concept analysis. *PLoS One, 9*(9). https://doi.org/10.1371/journal.pone.0107828

Skinner, T. C., Joensen, L., & Parkin, T. (2020). Twenty-five years of diabetes distress research. *Diabet. Med., 37*(3), 393–400. https://doi.org/10.1111/dme.14157

Smith, R. C., Marshall-Dorsey, A. A., Osborn, G. G., Shebroe, V., Lyles, J. S., Stoffelmayr, B. E., et al. (2000). Evidence-based guidelines for teaching patient-centered interviewing. *Patient Educ. Couns., 39*(1), 27–36. https://doi.org/10.1016/S0738-3991(99)00088-9

Speight, J., Conn, J., Dunning, T., & Skinner, T. C. (2012). Diabetes Australia position statement. A new language for diabetes: improving communications with and about people with diabetes. *Diabetes Res. Clin. Pract., 97*(3), 425–431. https://doi.org/10.1016/j.diabres.2012.03.015

Speight, J., Skinner, T., Dunning, T., Black, T., Kilov, G., Lee, C., Scibilia, R., & Johnson, G. (2021). Our language matters: improving communication with and about people with diabetes. A position statement by Diabetes Australia. *Diabetes Res. Clin. Pract., 173*, 108,655. https://doi.org/10.1016/j.diabres.2021.108655

Stalmeier, P. F. M., van Tol-Geerdink, J. J., van Lin, E. N. J. T., Schimmel, E., Huizenga, H., van Daal, W. A. J., & Leer, J. W. (2007). Doctors' and patients' preferences for participation and treatment in curative prostate cancer radiotherapy. *J. Clin. Oncol., 25*(21), 3096–3100. https://doi.org/10.1200/JCO.2006.07.4955

Vuohijoki, A., Mikkola, I., Jokelainen, J., Keinänen-Kiukaanniemi, S., Winell, K., Frittitta, L., Timonen, M., & Hagnäs, M. (2020). Implementation of personalized care plan for patients with Type 2 diabetes is associated with improvements in clinical outcomes: An observational real-world study. *Journal of Primary Care & Community Health, 11*, 1–7.

Walters, R., Leslie, S. J., Polson, R., Cusack, T., & Gorely, T. (2020). Establishing the efficacy of interventions to improve health literacy and health behaviours: a systematic review. *BMC Public Health, 20*(1), 1040. https://doi.org/10.1186/s12889-020-08991-0

Open Access   This chapter is licensed under the terms of the Creative Commons Attribution 4.0 International License (http://creativecommons.org/licenses/by/4.0/), which permits use, sharing, adaptation, distribution and reproduction in any medium or format, as long as you give appropriate credit to the original author(s) and the source, provide a link to the Creative Commons license and indicate if changes were made.

The images or other third party material in this chapter are included in the chapter's Creative Commons license, unless indicated otherwise in a credit line to the material. If material is not included in the chapter's Creative Commons license and your intended use is not permitted by statutory regulation or exceeds the permitted use, you will need to obtain permission directly from the copyright holder.

Chapter 6
Supplementary Online Resources for the Development of Behaviour Change Support Competencies

Gregor Štiglic, Katja Braam, Maria Beatriz Carmo, Luís Correia, Lucija Gosak, Mateja Lorber, Nuno Pimenta, and Ana Paula Cláudio

The Train4Health project addresses the challenge of the skill gap in behaviour change support through an innovative open-access educational package, comprising case studies, a massive open online course (MOOC) and a web application to simulate behaviour change support in persons with chronic disease.

These products were developed by an interdisciplinary team (nursing, pharmacy, sport science and informatics), in consultation with experts in behaviour change and interprofessional education.

G. Štiglic (✉)
Faculty of Health Sciences, University of Maribor, Maribor, Slovenia

Faculty of Electrical Engineering and Computer Science, University of Maribor, Maribor, Slovenia

Usher Institute, University of Edinburgh, Edinburgh, UK
e-mail: gregor.stiglic@um.si

K. Braam
Faculty of Health, Sports and Social Work, Inholland University of Applied Sciences, Haarlem, Netherlands

M. B. Carmo · L. Correia · A. Paula Cláudio
LASIGE, Faculdade de Ciências, Universidade de Lisboa, Lisbon, Portugal

L. Gosak · M. Lorber
Faculty of Health Sciences, University of Maribor, Maribor, Slovenia

N. Pimenta
Sport Sciences School of Rio Maior – Polytechnic Institute of Santarém, Rio Maior, Portugal

Interdisciplinary Centre for the Study of Human Performance, Faculty of Human Kinetics, University of Lisbon, Lisbon, Portugal

Centro de Investigação Interdisciplinar em Saúde, Instituto de Ciências da Saúde, Universidade Católica Portuguesa, Lisbon, Portugal

© The Author(s) 2023   113
M. P. Guerreiro et al. (eds.), *A Practical Guide on Behaviour Change Support for Self-Managing Chronic Disease*, https://doi.org/10.1007/978-3-031-20010-6_6

Overall, educational products provide learning activities aligned with the inter-professional competency framework and the learning outcomes. The latter were developed as part of a transnational curriculum, already described in Chap. 1.

Innovation was pursued by materialising ideas based on existing or newly produced knowledge, as listed below, leading to the generation of value:

- Knowledge on case studies and MOOC design, within-simulation feedback and post-simulation debriefing and theory on gamification, to guide development.
- Knowledge from European projects, such as Sim-Versity,[1] to ensure diversity and inclusion in case studies (e.g. ethnicity and sexual orientation), and WE4AHA,[2] to offer a holistic persons' profiles for case studies.
- Knowledge from behaviour change science, such as theories, models and a behaviour change taxonomy, to guide the content produced.
- Knowledge from transnational co-production with students and academic educators in nursing, pharmacy and sport science, to maximise the fit with perceived needs and preferred features.

Testing of the products and assessment of learning outcomes were conducted by different teams within the project, working collaboratively with developers, underpinned by a review on digital tools in behaviour change support education (Gosak et al., 2021). Testing was conducted iteratively, through improvement cycles, contributing to quality and alignment with users' preferences.

The MOOC provides an interactive format of the information presented in this book. Next, the two other products – case studies and the web application – are presented. Case studies are of particular interest for academic educators teaching behaviour change support in chronic diseases, while the web application offers flexibility to students independently pursuing training outside the classroom. This book informs the answers to case study questions and is a resource to post-simulation feedback in the web application, by providing readily accessible, peer-reviewed and up-to-date information. Both educational products direct learners to specific sections of this book and are open access (Box 6.1).

Box 6.1 Practical Aspects for Using the Case Study Toolkits and the Web Application

- Access the Train4Health website (https://www.train4health.eu/) for the most up-to-date releases and the full range of materials, in the "Resources" tab.
- Copyright under an open licence, which permits non-commercial no-cost access, re-use, re-purpose and redistribution by others, provided that the source is acknowledged (CC BY-NC-SA 4.0, https://creativecommons.org/licenses/by-nc-sa/4.0).

---

[1] Sim-Versity is an Erasmus plus project (https://sim-versity.eu), which aims to optimise patient safety by integrating cultural competence into simulation-based education of health professionals.

[2] Horizon 2020 project, WE4AHA (Widening the support for large-scale uptake of Digital Innovation for Active and Healthy Ageing), under the umbrella of which the Blueprint action created 12 personas (https://blueprint-personas.eu/)

6.1   Case Study Toolkits

Case studies are an instructional method that engages students in the discussion of specific situations, typically real-world examples, providing context and allowing students to learn in a controlled environment. Group work is a privileged form to reflect on the case and to collaboratively address questions that have no single right answer; the educator's role is facilitating decision-making and group work (Thistlethwaite et al., 2012).

Four case study toolkits were produced; each is composed of four components that work together, intending to provide the best teaching and learning experience:

- The **person's profile** presents the story of a person with one or more chronic diseases, unravelling health behaviour change opportunities. Each profile was primarily designed to support change in selected target behaviours, such as physical activity, smoking cessation and diet, with increasing complexity.
- **Learning outcomes and related resources**, linked with content topics, open-ended questions and suggested accompanying reading.
- **Assessment criteria**, for in-class group work, to be used by educators, and self-assessment criteria for students
- **Guidance for educators**, to aid the teaching process and classroom implementation.

The persons' profiles offer diversity, not only in terms of gender, age, ethnicity, sexual orientation and social and functional status but also geography (Fig. 6.1). There are versions available in languages other than English, some of which suffered cultural adaptation by the community of early adopters[3] of the project's educational products. Authenticity was pursued by relating each case to real life and through the consultation with persons living with chronic disease and other stakeholders.

Maria José (PT)   Nina (SL)   Liam (IE)   Luuk (NL)

Fig. 6.1 The four Train4Health persons' profiles, in increasing order of complexity

---

[3] A programme to engage educators or institutions in the early adoption of the T4H educational products

We suggest that in-class use of case studies should start with preparation by academic educators, requiring reading and getting familiar with the person's profile, selecting learning outcomes and questions, deciding whether pre-class and/or post-class assignments are warranted and planning in-class dynamics (e.g. group formation and reporting). Then students receive the person's profile and discuss the answers to the selected questions as group work, in-class.

Continuous improvement of case study components was based on internal peer-review and iterative tests with students and educators, focusing on aspects such as clarity, the perceived realism of person's profile, perceived value and overall opinion, through a self-administered online survey. In essence, testing was done in samples that did not experience classroom use of case studies and in those using case studies in class, across different geographies. In-class testing encompassed both face-to-face and online contexts, within and beyond the consortium, taking advantage of the Physical Activity and Lifestyle (PAL) network and the early adopter's programme. Globally, the case studies were tested by over 20 educators and 750 students, with overall positive feedback.

6.2   Web Application to Simulate Behaviour Change Support

The Train4Health web application offers a safe environment for training behaviour change support, in which users play the role of a health or another professional and interact with 2D virtual humans, playing the role of persons with chronic disease. The web application allows convenient access via web browsers in a range of devices (smart phones, tablets, laptops and desktops) without the installation of a specific application.

Simulation consists of interactive experiences with the four animated persons' profiles, depicted in Fig. 6.1. These virtual humans are capable of speaking through a synthetic voice (currently in English only) and express different facial expressions. Subtitles shown in the interface ensure inclusiveness for users with hearing impairment or less proficiency in English.

The web application is composed of three key menus: simulations, performance and resources.

The simulation experience starts with the case of Maria José, which is the less complex case (Fig. 6.1). The others are sequentially unlocked as sessions with lower levels of difficulty are completed, allowing a stepwise learning. Users are offered the opportunity to train both brief and long interventions.

Before starting each session, learning outcomes are presented. Then, the user engages in dialogue with the selected virtual human through buttons presenting two options, one of which is considered more correct (Fig. 6.2). Selecting the more correct option determines progression in the dialogue; choosing the less correct option prompts performance-based feedback and then directs the user to the conversational thread.

Fig. 6.2 Interface with Nina's dialogue (SimSoft2.0)

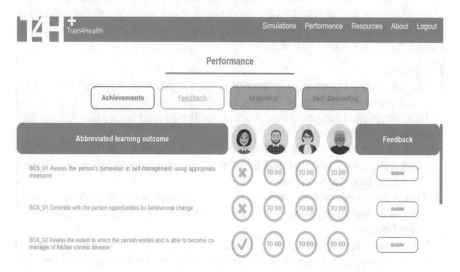

Fig. 6.3 Interface with user performance in relation to learning outcomes (SimSoft2.0)

At the end of the session, the user gets feedback in relation to each learning outcome plus a performance score. Feedback directs the user for resources to improve learning, where needed, via the "Resources menu". Finally, the user is invited for a self-debriefing session via Gibbs' reflective cycle (MacKenna et al., 2021).

The performance menu displays selected gamification features, such as points, statistics and acknowledgement through badges. The Self-Determination Theory enabled an understanding of how gamification enhances engagement and motivation while helping to avoid pitfalls in implementation (Rutledge et al., 2018). Assessment of learning outcomes in each session and across sessions is computed through algorithms informed by the choice of answer options. A dashboard with the user performance per learning outcomes across scenarios is also depicted (Fig. 6.3), accompanied by feedback and feedforward for resources.

The **Resources** menu encompasses information on the ABCD approach for structuring behaviour change support (detailed in Chap. 5), a list of core behaviour change techniques for the self-management of chronic disease, a glossary (Guerreiro et al., 2021) and a hyperlink to this book.

The web application was tested iteratively with students and educators. In the first iteration, usability was evaluated using a multi-method approach, comprising the system usability scale (Brooke, 1996) coupled to additional questions, previously piloted plus task performance with a think-aloud protocol (Silva et al., 2021). Data were collected in online individual sessions with a sample of 12 students, 15 educators and 3 researchers, from 4 nationalities. Almost all participants were able to complete the full set of proposed tasks; only about 1% of the tasks were not completed. The overall experience with the simulation software was scored 4.5 in a 5-point Likert scale (1, bad; 5, excellent) whilst the average result of the System Usability Scale (SUS) was 85.7 (scale 0–100), which can be classified as an excellent score (Bangor et al., 2009). After minor adjustments, subsequent iterations explored specific features of the web application and its usability in real use context. This means that participants had the opportunity to use the software individually at their own pace, followed by data collection through a self-administered online questionnaire, which included the SUS.

Key Points
- Ideas leading to innovative behaviour change support education products were materialised by resorting to existing knowledge from diverse fields (e.g. education, behaviour change science, transnational co-production with students and academic educators).
- Case studies and the web application were developed by an interdisciplinary team, guided by the Train4Health interprofessional competency framework and associated learning outcomes.
- This coherent educational package is expected to be useful to a broad range of groups, including academic educators, students and professionals of health and related fields.
- These educational products are available under an open licence that permits no-cost access, re-use, re-purpose and redistribution by others, provided the source is acknowledged.

References

Bangor, A., Kortum, P., & Miller, J. (2009). Determining what individual SUS scores mean: adding an adjective rating scale. *J. Usability Stud., 4*(3), 114–123.

Brooke, J. (1996). SUS: a "quick and dirty" usability scale. In P. W. Jordan, B. Thomas, B. A. Weerdmeester, & I. L. McClelland (Eds.), *Usability Evaluation in Industry (Pp. 189–194)*. Taylor & Francis.

Gosak, L., Štiglic, G., Budler, L. C., Félix, I. B., Braam, K., Fijačko, N., Guerreiro, M. P., & Lorber, M. (2021). Digital tools in behavior change support education in health and other students: a systematic review. *Healthcare, 10*(1). https://doi.org/10.3390/healthcare10010001

Guerreiro, M. P., Strawbridge, J., Cavaco, A. M., Félix, I. B., Marques, M. M., & Cadogan, C. (2021). Development of a European competency framework for health and other professionals to support behaviour change in persons self-managing chronic disease. *BMC Med. Educ., 21*(1), 1–14. https://doi.org/10.1186/s12909-021-02720-w

MacKenna, V., Díaz, D. A., Chase, S. K., Boden, C. J., & Loerzel, V. (2021). Self-debriefing in healthcare simulation: an integrative literature review. *Nurse Educ. Today, 102,* 104907. https://doi.org/10.1016/j.nedt.2021.104907

Rutledge, C., Walsh, C. M., Swinger, N., Auerbach, M., Castro, D., Dewan, M., Khattab, M., Rake, A., Harwayne-Gidansky, I., Raymond, T. T., Maa, T., Chang, T. P., & Quality Cardiopulmonary Resuscitation (QCPR) leaderboard investigators of the International Network for Simulation-based Pediatric Innovation, Research, and Education (INSPIRE). (2018). Gamification in action: theoretical and practical considerations for medical educators. *Acad. Med., 93*(7), 1014–1020. https://doi.org/10.1097/ACM.0000000000002183

Silva, A. G., Caravau, H., Martins, A., Almeida, A. M. P., Silva, T., Ribeiro, Ó., et al. (2021). Procedures of user-centered usability assessment for digital solutions: scoping review of reviews reporting on digital solutions relevant for older adults. *JMIR Hum. Factors, 8*(1), 1–14. https://doi.org/10.2196/22774

Thistlethwaite, J. E., Davies, D., Ekeocha, S., Kidd, J. M., MacDougall, C., Matthews, P., Purkis, J., & Clay, D. (2012). The effectiveness of case-based learning in health professional education. A BEME systematic review: BEME Guide No. 23. *Med. Teach., 34*(6), e421–e444. https://doi.org/10.3109/0142159X.2012.680939

Open Access This chapter is licensed under the terms of the Creative Commons Attribution 4.0 International License (http://creativecommons.org/licenses/by/4.0/), which permits use, sharing, adaptation, distribution and reproduction in any medium or format, as long as you give appropriate credit to the original author(s) and the source, provide a link to the Creative Commons license and indicate if changes were made.

The images or other third party material in this chapter are included in the chapter's Creative Commons license, unless indicated otherwise in a credit line to the material. If material is not included in the chapter's Creative Commons license and your intended use is not permitted by statutory regulation or exceeds the permitted use, you will need to obtain permission directly from the copyright holder.

Index

© The Editor(s) (if applicable) and The Author(s) 2023
M. P. Guerreiro et al. (eds.), *A Practical Guide on Behaviour Change Support
for Self-Managing Chronic Disease*, https://doi.org/10.1007/978-3-031-20010-6

Printed in the United States
by Baker & Taylor Publisher Services